Series

Putting **Essential Understanding** of

Ratios and Proportions into Practice

in Grades
6–8

Travis A. Olson
University of Nevada, Las Vegas
Las Vegas, Nevada

Melfried Olson
University of Hawaii at Manoa
Honolulu, Hawaii

Hannah Slovin
University of Hawaii at Manoa
Honolulu, Hawaii

Hannah Slovin
Volume Editor
University of Hawaii at Manoa
Honolulu, Hawaii

Barbara J. Dougherty
Series Editor
University of Missouri
Columbia, Missouri

1906 Association Drive, Reston, VA 20191-1502
(703) 620-9840; (800) 235-7566; www.nctm.org

Library of Congress Cataloging-in-Publication Data

Olson, Travis, 1978- author.
Putting essential understanding of ratios and proportions into practice in grades 6–8 / Travis Olson, University of Nevada, Las Vegas, Las Vegas, Nevada, Melfried Olson, University of Hawaii at Manoa, Honolulu, Hawaii, Hannah Slovin, University of Hawaii at Manoa Honolulu, Hawaii ; volume editor, Hannah Slovin, University of Hawaii at Manoa, Honolulu, Hawaii.
pages cm
Includes bibliographical references.
ISBN 978-0-87353-717-9
1. Ratio and proportion—Study and teaching (Middle school) I. Olson, Melfried, author. II. Slovin, Hannah, author, editor. III. Title.
QA117.047 2015
513.2'4—dc23

2014034046

The National Council of Teachers of Mathematics is the public voice of mathematics education, providing vision, leadership, and professional development to support teachers in ensuring equitable mathematics learning of the highest quality for all students.

Printed in the United States of America

Contents

Foreword vii

Preface ix

Introduction 1

Pedagogical Content Knowledge 1

Model of Teacher Knowledge 3

Characteristics of Tasks 7

Types of Questions 9

Conclusion 10

Chapter 1
Attending to Invariance in Ratios and Proportions 11

Working toward the Big Idea of Ratios and Proportions 12

Summarizing Pedagogical Content Knowledge to Support the Big Idea 22

Knowledge of learners 22

Knowledge of curriculum 23

Knowledge of instructional strategies 26

Knowledge of assessment 28

Conclusion 28

Chapter 2
Using Ratios to Measure Attributes 31

Working toward Essential Understanding 3 33

Using direct measurements of an attribute to compare objects 33

Moving to relational attributes: Probability and speed 39

Summarizing Pedagogical Content Knowledge to Support Essential Understanding 3 48

Knowledge of learners 48

Knowledge of curriculum 49

Knowledge of instructional strategies 50

Knowledge of assessment 50

Conclusion 51

Chapter 3

Focusing on Equivalent Ratios 53

Working toward Essential Understanding 7 54

Summarizing Pedagogical Content Knowledge to Support Essential Understanding 7 66

Knowledge of learners 66

Knowledge of curriculum 67

Knowledge of instructional strategies 67

Knowledge of assessment 68

Conclusion 69

Chapter 4

Reasoning about Rates 71

Working toward Essential Understanding 8 72

Summarizing Pedagogical Content Knowledge to Support Essential Understanding 8 80

Knowledge of learners 80

Knowledge of curriculum 81

Knowledge of instructional strategies 82

Knowledge of assessment 83

Conclusion 84

Chapter 5
Generalizing Reasoning to Solve Proportion Problems ... 85

Working toward Essential Understanding 9 ... 85

Summarizing Pedagogical Content Knowledge to Support Essential Understanding 9 ... 98

Knowledge of learners ... 98

Knowledge of curriculum ... 99

Knowledge of instructional strategies ... 99

Knowledge of assessment ... 100

Conclusion ... 101

Chapter 6
Looking Back and Looking Ahead with Ratios and Proportions ... 103

Building Foundations for Ratios and Proportions in Grades 3–5 ... 103

Extending Understanding of Ratios and Proportions in Grades 9–12 ... 112

Conclusion ... 117

Appendix 1
The Big Idea and Essential Understandings for Ratios, Proportions, and Proportional Reasoning ... 121

Appendix 2
Resources for Teachers ... 123

Appendix 3
Tasks ... 131

Sands of Time ... 132

Pizza Presto: Which Chef Is Faster? ... 133

Measure for Measure, How Long Was That Board? ... 134

Loads of Luscious Lilikoi ... 135

Off the Top of Your Head, Which Is Larger? 136

How Many Can You Find? 137

References 139

Accompanying Materials at More4U

Appendix 1
The Big Idea and Essential Understandings for Ratios, Proportions, and Proportional Reasoning

Appendix 2
Resources for Teachers

Appendix 3
Tasks

Sands of Time

Punching Up the Recipe

You Tell Me, Which Is Larger?

Red Gumball or White? That Is the Question!

Pizza Presto: Which Chef Is Faster?

Are They Equivalent?

Measure for Measure, How Long Was That Board?

Loads of Luscious Lilikoi

Off the Top of Your Head, Which Is Larger?

Brainstorming about the Better Buy

Relating Reciprocals to 1

All the Same When Simplified

How Many Can You Find?

Foreword

Teaching mathematics in prekindergarten–grade 12 requires knowledge of mathematical content and developmentally appropriate pedagogical knowledge to provide students with experiences that help them learn mathematics with understanding, while they reason about and make sense of the ideas that they encounter.

In 2010 the National Council of Teachers of Mathematics (NCTM) published the first book in the Essential Understanding Series, focusing on topics that are critical to the mathematical development of students but often difficult to teach. Written to deepen teachers' understanding of key mathematical ideas and to examine those ideas in multiple ways, the Essential Understanding Series was designed to fill in gaps and extend teachers' understanding by providing a detailed survey of the big ideas and the essential understandings related to particular topics in mathematics.

The Putting Essential Understanding into Practice Series builds on the Essential Understanding Series by extending the focus to classroom practice. These books center on the pedagogical knowledge that teachers must have to help students master the big ideas and essential understandings at developmentally appropriate levels.

To help students develop deeper understanding, teachers must have skills that go beyond knowledge of content. The authors demonstrate that for teachers–

- understanding student misconceptions is critical and helps in planning instruction;
- knowing the mathematical content is not enough–understanding student learning and knowing different ways of teaching a topic are indispensable;
- constructing a task is important because the way in which a task is constructed can aid in mediating or negotiating student misconceptions by providing opportunities to identify those misconceptions and determine how to address them.

Through detailed analysis of samples of student work, emphasis on the need to understand student thinking, suggestions for follow-up tasks with the potential to move students forward, and ideas for assessment, the Putting Essential Understanding into Practice Series demonstrates best practice for developing students' understanding of mathematics.

The ideas and understandings that the Putting Essential Understanding into Practice Series highlights for student mastery are also embodied in the Common Core State

Standards for Mathematics, and connections with these new standards are noted throughout each book.

On behalf of the Board of Directors of NCTM, I offer sincere thanks to everyone who has helped to make this new series possible. Special thanks go to Barbara J. Dougherty for her leadership as series editor and to all the authors for their work on the Putting Essential Understanding into Practice Series. I join the project team in welcoming you to this special series and extending best wishes for your ongoing enjoyment—and for the continuing benefits for you and your students—as you explore Putting Essential Understanding into Practice!

Linda M. Gojak
President, 2012–2014
National Council of Teachers of Mathematics

Preface

The Putting Essential Understanding into Practice Series explores the teaching of mathematics topics in K–grade 12 that are difficult to learn and to teach. Each volume in this series focuses on specific content from one volume in NCTM's Essential Understanding Series and links it to ways in which those ideas can be taught successfully in the classroom.

Thus, this series builds on the earlier series, which aimed to present the mathematics that teachers need to know and understand well to teach challenging topics successfully to their students. Each of the earlier books identified and examined the big ideas related to the topic, as well as the "essential understandings"—the associated smaller, and often more concrete, concepts that compose each big idea.

Taking the next step, the Putting Essential Understanding into Practice Series shifts the focus to the specialized pedagogical knowledge that teachers need to teach those big ideas and essential understandings effectively in their classrooms. The Introduction to each volume details the nature of the complex, substantive knowledge that is the focus of these books—*pedagogical content knowledge*. For the topics explored in these books, this knowledge is both student centered and focused on teaching mathematics through problem solving.

Each book then puts big ideas and essential understandings related to the topic under a high-powered teaching lens, showing in fine detail how they might be presented, developed, and assessed in the classroom. Specific tasks, classroom vignettes, and samples of student work serve to illustrate possible ways of introducing students to the ideas in ways that will enable students not only to make sense of them now but also to build on them in the future. Items for readers' reflection appear throughout and offer teachers additional opportunities for professional development.

The final chapter of each book looks at earlier and later instruction on the topic. A look back highlights effective teaching that lays the earlier foundations that students are expected to bring to the current grades, where they solidify and build on previous learning. A look ahead reveals how high-quality teaching can expand students' understanding when they move to more advanced levels.

Each volume in the Putting Essential Understanding into Practice Series also includes three appendixes to extend and enrich readers' experiences and possibilities for using the book. The appendixes list the big ideas and essential understandings related to the topic, detail resources for teachers, and present tasks discussed in the book. These materials are also available to readers online at the More4U website, where Appendix 3 includes additional tasks in a format to facilitate hands-on work with students. Readers can gain online access to each book's More4U materials by going to www.nctm.org/more4u and entering the code that appears on the title page. They can then print out these materials for personal or classroom use.

Because the topics chosen for both the earlier Essential Understanding Series and this successor series represent areas of mathematics that are widely regarded as challenging to teach and to learn, we believe that these books fill a tangible need for teachers. We hope that as you move through the tasks and consider the associated classroom implementations, you will find a variety of ideas to support your teaching and your students' learning.

Acknowledgments from the Authors

We wish to express our thanks for the various contributions to the development of the manuscript of this book from the following: students at University Laboratory School, Honolulu, Hawaii; Brendan Brennan, University Laboratory School, Honolulu, Hawaii; Stephanie Capen, Honolulu, Hawaii; and Maryam Abhari, Honolulu, Hawaii.

Introduction

Shulman (1986, 1987) identified seven knowledge bases that influence teaching:

1. Content knowledge
2. General pedagogical knowledge
3. Curriculum knowledge
4. Knowledge of learners and their characteristics
5. Knowledge of educational contexts
6. Knowledge of educational ends, purposes, and values
7. Pedagogical content knowledge

The specialized content knowledge that you use to transform your understanding of mathematics content into ways of teaching is what Shulman identified as item 7 on this list–*pedagogical content knowledge* (Shulman 1986). This is the knowledge that is the focus of this book–and all the volumes in the Putting Essential Understanding into Practice Series.

Pedagogical Content Knowledge

In mathematics teaching, pedagogical content knowledge includes at least four indispensable components:

1. Knowledge of curriculum for mathematics
2. Knowledge of assessments for mathematics
3. Knowledge of instructional strategies for mathematics
4. Knowledge of student understanding of mathematics (Magnusson, Krajcik, and Borko 1999)

These four components are linked in significant ways to the content that you teach.

Even though it is important for you to consider how to structure lessons, deciding what group and class management techniques you will use, how you will allocate time, and what will be the general flow of the lesson, Shulman (1986) noted that it is even more important to consider *what* is taught and the *way* in which it is taught. Every day, you make at least five essential decisions as you determine–

1. which explanations to offer (or not);
2. which representations of the mathematics to use;
3. what types of questions to ask;
4. what depth to expect in responses from students to the questions posed; and
5. how to deal with students' misunderstandings when these become evident in their responses.

Your pedagogical content knowledge is the unique blending of your content expertise and your skill in pedagogy to create a knowledge base that allows you to make robust instructional decisions. Shulman (1986, p. 9) defined pedagogical content knowledge as "a second kind of content knowledge..., which goes beyond knowledge of the subject matter per se to the dimension of subject matter knowledge *for teaching*." He explained further:

> Pedagogical content knowledge also includes an understanding of what makes the learning of specific topics easy or difficult: the conceptions and preconceptions that students of different ages and backgrounds bring with them to the learning of those most frequently taught topics and lessons. (p. 9)

If you consider the five decision areas identified at the top of the page, you will note that each of these requires knowledge of the mathematical content and the associated pedagogy. For example, teaching ratios and proportions requires that you understand the difference between part-whole and part-part comparisons. Your knowledge of ratios and proportions can help you craft tasks and questions that provide counterexamples and ways to guide your students in seeing connections across multiple number systems. As you establish the content, complete with learning goals, you then need to consider how to move your students from their initial understandings to deeper ones, building rich connections along the way.

The instructional sequence that you design to meet student learning goals has to take into consideration the misconceptions and misunderstandings that you might expect to encounter (along with the strategies that you expect to use to negotiate them), your expectation of the level of difficulty of the topic for your students, the progression of experiences in which your students will engage, appropriate collections of representations for the content, and relationships between and among ratios and proportions and other topics.

Model of Teacher Knowledge

Grossman (1990) extended Shulman's ideas to create a model of teacher knowledge with four domains (see fig. 0.1):

1. Subject-matter knowledge
2. General pedagogical knowledge
3. Pedagogical content knowledge
4. Knowledge of context

Subject-matter knowledge includes mathematical facts, concepts, rules, and relationships among concepts. Your understanding of the mathematics affects the way in which you teach the content—the ideas that you emphasize, the ones that you do not, particular algorithms that you use, and so on (Hill, Rowan, and Ball 2005).

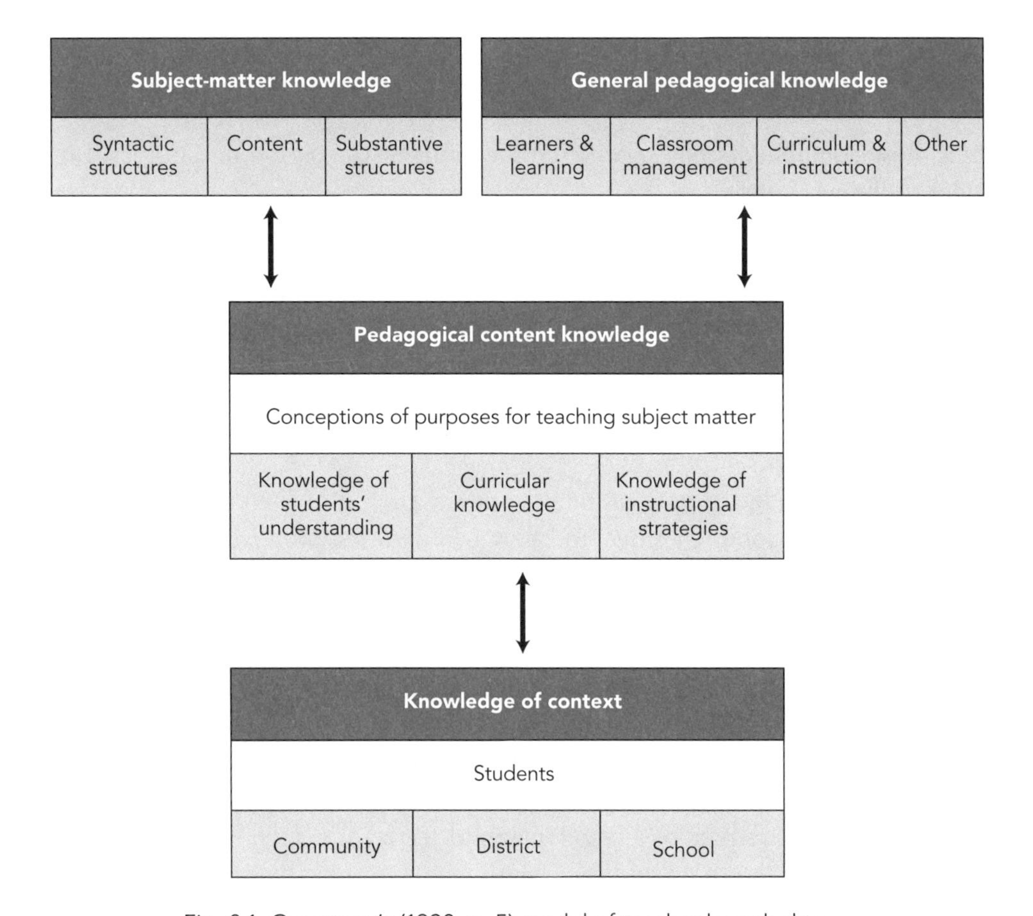

Fig. 0.1. Grossman's (1990, p. 5) model of teacher knowledge

Your pedagogical knowledge relates to the general knowledge, beliefs, and skills that you possess about instructional practices. These include specific instructional strategies that you use, the amount of wait time that you allow for students' responses to questions or tasks, classroom management techniques that you use for setting expectations and organizing students, and your grouping techniques, which might include having your students work individually or cooperatively or collaboratively, in groups or pairs. As Grossman's model indicates, your understanding and interpretation of the environment of your school, district, and community can also have an impact on the way in which you teach a topic.

Note that pedagogical content knowledge has four aspects, or components, in Grossman's (1990) model:

1. Conceptions of purposes for teaching
2. Knowledge of students' understanding
3. Knowledge of curriculum
4. Knowledge of instructional strategies

Each of these components has specific connections to the classroom. It is useful to consider each one in turn.

First, when you think about the goals that you want to establish for your instruction, you are focusing on your conceptions of the purposes for teaching. This is a broad category but an important one because the goals that you set will define learning outcomes for your students. These conceptions influence the other three components of pedagogical content knowledge. Hence, they appropriately occupy their overarching position in the model.

Second, your knowledge of your students' understanding of the mathematics content is central to good teaching. To know what your students understand, you must focus on both their conceptions and their misconceptions. As teachers, we all recognize that students develop naïve understandings that may or may not be immediately evident to us in their work or discourse. These can become deep-rooted misconceptions that are not simply errors that students make. Misconceptions may include incorrect generalizations that students have developed, such as thinking that they can model ratios with sets of discrete objects that can be combined to show the addition of ratios. These generalizations may even be predictable notions that students exhibit as part of a developmental trajectory, such as thinking that equivalent ratios are based on additive relationships rather than multiplicative ones.

Part of your responsibility as a teacher is to present tasks or to ask questions that can bring misconceptions to the forefront. Once you become aware of misconceptions in students' thinking, you then have to determine the next instructional steps. The mathematical ideas presented in this volume focus on common misconceptions that students form in relation to a specific topic—ratios and proportions in grades 6–8. This book shows how the type of task selected and the sequencing of carefully developed questions can bring the misconceptions to light, as well as how particular teachers took the next instructional steps to challenge their students' misconceptions.

Third, curricular knowledge for mathematics includes multiple areas. Your teaching may be guided by a set of standards such as the Common Core State Standards for Mathematics (CCSSM; National Governors' Association Center for Best Practices and Council of Chief State School Officers 2010) or other provincial, state, or local standards. You may in fact use these standards as the learning outcomes for your students. Your textbook is another source that may influence your instruction. With any textbook also comes a particular philosophical view of mathematics, mathematics teaching, and student learning. Your awareness and understanding of the curricular perspectives related to the choice of standards and the selection of a textbook can help to determine how you actually enact your curriculum. Moreover, your district or school may have a pacing guide that influences your delivery of the curriculum. In this book, we can focus only on the alignment of the topics presented with broader curricular perspectives, such as CCSSM. However, your own understanding of and expertise with your other curricular resources, coupled with the parameters defined by the expected student outcomes from standards documents, can provide the specificity that you need for your classroom.

In addition to your day-to-day instructional decisions, you make daily decisions about which tasks from curricular materials you can use without adaptation, which tasks you will need to adapt, and which tasks you will need to create on your own. Once you select or develop meaningful, high-quality tasks and use them in your mathematics lesson, you have launched what Yinger (1988) called "a three-way conversation between teacher, student, and problem" (p. 86). This process is not simple—it is complex because how students respond to the problem or task is directly linked to your next instructional move. That means that you have to plan multiple instructional paths to choose among as students respond to those tasks.

Knowledge of the curriculum goes beyond the curricular materials that you use. You also consider the mathematical knowledge that students bring with them from grade 5 and what they should learn by the end of grade 8. The way in which you teach a foundational concept or skill has an impact on the way in which students will interact with and learn later related content. For example, the types of representations that you include in your introduction of ratios and proportions are the ones that your students will use to evaluate other representations and ideas in later grades.

Fourth, knowledge of instructional strategies is essential to pedagogical content knowledge. Having a wide array of instructional strategies for teaching mathematics is central to effective teaching and learning. Instructional strategies, along with knowledge of the curriculum, may include the selection of mathematical tasks, together with the way in which those tasks will be enacted in the classroom. Instructional strategies may also include the way in which the mathematical content will be structured for students. You may have very specific ways of thinking about how you will structure your presentation of a mathematical idea—not only how you will sequence the introduction and development of the idea, but also how you will present that idea to your students. Which examples should you select, and which questions should you ask? What representations should you use? Your knowledge of instructional strategies, coupled with your knowledge of your curriculum, permits you to align the selected mathematical tasks closely with the way in which your students perform those tasks in your classroom.

The instructional approach in this volume combines a student-centered perspective with an approach to mathematics through problem solving. A student-centered approach is characterized by a shared focus on student and teacher conversations, including interactions among students. Students who learn through such an approach are active in the learning process and develop ways of evaluating their own work and one another's in concert with the teacher's evaluation.

Teaching through problem solving makes tasks or problems the core of mathematics teaching and learning. The introduction to a new topic consists of a task that students work through, drawing on their previous knowledge while connecting it with new ideas. After students have explored the introductory task (or tasks), their consideration of solution methods, the uniqueness or multiplicity of solutions, and extensions of the task create rich opportunities for discussion and the development of specific mathematical concepts and skills.

By combining the two approaches, teachers create a dynamic, interactive, and engaging classroom environment for their students. This type of environment promotes the ability of students to demonstrate CCSSM's Standards for Mathematical Practice while learning the mathematics at a deep level.

The chapters that follow will show that instructional sequences embed all the characteristics of knowledge of instructional strategies that Grossman (1990) identifies. One component that is not explicit in Grossman's model but is included in a model developed by Magnusson, Krajcik, and Borko (1999) is the knowledge of assessment. Your knowledge of assessment in mathematics plays an important role in guiding your instructional decision-making process.

There are different types of assessments, each of which can influence the evidence that you collect as well as your view of what students know (or don't know) and how they know what they do. Your interpretation of what students know is also related to your view of what constitutes "knowing" in mathematics. As you examine the tasks, classroom vignettes, and samples of student work in this volume, you will notice that teacher questioning permits formative assessment that supplies information that spans both conceptual and procedural aspects of understanding. *Formative assessment*, as this book uses the term, refers to an appraisal that occurs during an instructional segment, with the aim of adjusting instruction to meet the needs of students more effectively (Popham 2006). Formative assessment does not always require a paper-and-pencil product but may include questions that you ask or tasks that students complete during class.

The information that you gain from student responses can provide you with feedback that guides the instructional flow, while giving you a sense of how deeply (or superficially) your students understand a particular idea—or whether they hold a misconception that is blocking their progress. As you monitor your students' development of rich understanding, you can continually compare their responses with your expectations and then adapt your instructional plans to accommodate their current levels of development. Wiliam (2007, p. 1054) described this interaction between teacher expectations and student performance in the following way:

> It is therefore about assessment functioning as a bridge between teaching and learning, helping teachers collect evidence about student achievement in order to adjust instruction to better meet student learning needs, in real time.

Wiliam notes that for teachers to get the best information about student understandings, they have to know how to facilitate substantive class discussions, choose tasks that include opportunities for students to demonstrate their learning, and employ robust and effective questioning strategies. From these strategies, you must then interpret student responses and scaffold their learning to help them progress to more complex ideas.

Characteristics of Tasks

The type of task that is presented to students is very important. Tasks that focus only on procedural aspects may not help students learn a mathematical idea deeply.

Superficial learning may result in students forgetting easily, requiring reteaching, and potentially affecting how they understand mathematical ideas that they encounter in the future. Thus, the tasks selected for inclusion in this volume emphasize deep learning of significant mathematical ideas. These rich, "high-quality" tasks have the power to create a foundation for more sophisticated ideas and support an understanding that goes beyond "how" to "why." Figure 0.2 identifies the characteristics of a high-quality task.

As you move through this volume, you will notice that it sequences tasks for each mathematical idea so that they provide a cohesive and connected approach to the identified concept. The tasks build on one another to ensure that each student's thinking becomes increasingly sophisticated, progressing from a novice's view of the content to a perspective that is closer to that of an expert. We hope that you will find the tasks useful in your own classes.

A high-quality task has the following characteristics:
Aligns with relevant mathematics content standard(s)
Encourages the use of multiple representations
Provides opportunities for students to develop and demonstrate the mathematical practices
Involves students in an inquiry-oriented or exploratory approach
Allows entry to the mathematics at a low level (all students can begin the task) but also has a high ceiling (some students can extend the activity to higher-level activities)
Connects previous knowledge to new learning
Allows for multiple solution approaches and strategies
Engages students in explaining the meaning of the result
Includes a relevant and interesting context

Fig. 0.2. Characteristics of a high-quality task

Types of Questions

The questions that you pose to your students in conjunction with a high-quality task may at times cause them to confront ideas that are at variance with or directly contradictory to their own beliefs. The state of mind that students then find themselves in is called *cognitive dissonance,* which is not a comfortable state for students—or, on occasion, for the teacher. The tasks in this book are structured in a way that forces students to deal with two conflicting ideas. However, it is through the process of negotiating the contradictions that students come to know the content much more deeply. How the teacher handles this negotiation determines student learning.

You can pose three types of questions to support your students' process of working with and sorting out conflicting ideas. These questions are characterized by their potential to encourage reversibility, flexibility, and generalization in students' thinking (Dougherty 2001). All three types of questions require more than a one-word or one-number answer. Reversibility questions are those that have the capacity to change the direction of students' thinking. They often give students the solution and require them to create the corresponding problem. A flexibility question can be one of two types: it can ask students to solve a problem in more than one way, or it can ask them to compare and contrast two or more problems or determine the relationship between or among concepts and skills. Generalization questions also come in two types: they ask students to look at multiple examples or cases and find a pattern or make observations, or they ask them to create a specific example of a rule, conjecture, or pattern. Figure 0.3 provides examples of reversibility, flexibility, and generalization questions related to ratios and proportions.

Type of question	Example
Reversibility question	Jess found three fractions close to, but not equal to, $^1/_2$. What could the fractions be?
Flexibility question	If the ratio of girls to boys in the class is 3:2, how many boys and girls could be in the class? Give 3 examples.
Flexibility question	What ratio is equivalent to 4:5? Show at least two ways to find your ratio.
Generalization question	What are three true statements that could describe the set of fractions $^{43}/_{84}$, $^{43}/_{83}$, and $^{43}/_{85}$? Write a fraction that would (or would not) fit in this set. Explain why you chose that fraction.
Generalization question	What is an example of a ratio that has a unit rate of 1:6?

Fig. 0.3. Examples of reversibility, flexibility, and generalization questions

Conclusion

The Introduction has provided a brief overview of the nature of—and necessity for—pedagogical content knowledge. This knowledge, which you use in your classroom every day, is the indispensable medium through which you transmit your understanding of the big ideas of the mathematics to your students. It determines your selection of appropriate, high-quality tasks and enables you to ask the types of questions that will not only move your students forward in their understanding but also allow you to determine the depth of that understanding.

The chapters that follow describe important ideas related to learners, curricular goals, instructional strategies, and assessment that can assist you in transforming your students' knowledge into formal mathematical ideas related to ratios and proportions. These chapters provide specific examples of mathematical tasks and student thinking for you to analyze to develop your pedagogical content knowledge for teaching these topics in grades 6–8 or to give you ideas to help other colleagues develop this knowledge. You will also see how to bring together and interweave your knowledge of learners, curriculum, instructional strategies, and assessment to support your students in grasping the big ideas and essential understandings and using them to build more sophisticated knowledge.

Students in grades 6–8 have already had some experiences that affect their initial understanding of ratios and proportions. Furthermore, they have developed some ideas about these topics at earlier levels. Students in elementary classrooms frequently demonstrate understanding of mathematical ideas related to ratios and proportions in a particular context or in connection with a specific picture or drawing. Yet, in other situations, these same students do not demonstrate that same understanding. As their teacher, you must understand the ideas that they have developed about ratios and proportions in their prior experiences so you can extend this knowledge and see whether or how it differs from the formal mathematical knowledge that they need to be successful in reasoning with or applying ratios and proportions. You have the important responsibility of assessing their current knowledge related to the big idea of ratios and proportions as well as their understanding of various representations of these topics and their power and limitations. Your understanding will facilitate and reinforce your instructional decisions. Teaching the big idea and helping students develop essential understandings related to ratios and proportions is obviously a very challenging and complex task.

into practice

Chapter 1 Attending to Invariance in Ratios and Proportions

Big Idea

When two quantities are related proportionally, the ratio of one quantity to the other is invariant as the numerical values of both quantities change by the same factor.

One big idea underlies ratios and proportions, as identified in *Developing Essential Understanding of Ratios, Proportions, and Proportional Reasoning for Teaching Mathematics in Grades 6–8* (Lobato and Ellis 2010). This is the idea of invariance among quantities in ratios and proportional relationships. A critical aspect of students' understanding of ratios and proportions is the notion that as one quantity in a proportional relationship changes, so does the other quantity–and by the same factor.

For at least the past two centuries, proportional relationships have been regarded as a critical topic for students to understand, and proportional reasoning as an essential skill for them to develop, in school mathematics. Cohen (2003) notes that books used by students in the nineteenth century often ended with the "rule of three." This was a procedural mechanism enabling students to attend to the invariance in a proportional relationship without understanding that invariance conceptually. Stated simply, the rule of three was an algorithm that "instructed the student to set down the three known numbers in a particular order, and then multiply two and divide the product by the third" (Cohen 2003, p. 51). The following is an example of a problem to which students might apply the rule of three (Carter et al. 2013, p. 41):

> Cans of corn are on sale at 10 for $4. Find the cost of 15 cans.

Determining the ratio of 15 to an unknown dollar amount that is the same as the ratio of 10 to $4 solves the problem. That is, by setting up an equation of fractions to indicate equivalent ratios,

$$\frac{10}{4} = \frac{15}{x},$$

and solving for x, students find that the cost of 15 cans is $6.

This problem and others like it would fall under what Nicolas Pike (1809) characterized as the "rule of three direct": "If more require more, or less require less, the question belongs to the Rule of Three Direct. But if more require less, or less require more, it belongs to the Rule of Three Inverse" (p. 101).

In contrast to this rule-driven problem solving, the approach of Warren Colburn's early nineteenth-century textbooks more closely reflects that of recent NCTM publications. For example, one of Colburn's textbooks presented the following problem (1821, p. 44):

> A man had forty-two barrels of flour, and sold two sevenths of it for six dollars a barrel; how much did it come to?

Cohen (2003) notes, "Under the old arithmetic texts, this was a problem for the rule of three calling for careful written work, but Colburn expected students could reason it out without benefit of formula or paper" (p. 58). Cohen remarks that textbook authors such as Colburn (1826, p. 7) recognized the potential for the procedural use of the rule of three to interfere with the development of students' conceptual understanding of ratios, proportions, and proportional relationships:

> Those who understand the principles sufficiently to comprehend the nature of the rule of three, can do much better without it than with it, for when it is used, it obscures, rather than illustrates, the subject to which it is applied.

Thus, as early as 1826, U.S. textbook authors were drawing attention to proportional reasoning and the role of ratios and proportions in developing students' understanding of underlying concepts of invariance and scalability.

Along with building on the Big Idea and the related essential understandings outlined by Lobato and Ellis (2010), we hope to illustrate ways of working with invariance, scaling, and rates to develop students' understandings of ratios, proportions, and proportional relationships. This chapter focuses on invariance, the Big Idea of ratios and proportions, to establish a foundation for the chapters that follow.

Working toward the Big Idea of Ratios and Proportions

Providing your students with opportunities to engage in rich mathematical discussions engages you in two activities:

- Selecting, adapting, and designing tasks that will allow you to interpret your students' work

- Making thoughtful instructional decisions to build on your students' understanding

This work requires you, the teacher, to constantly challenge yourself to develop the specialized knowledge that will allow you to provide such opportunities for your students.

Developing this specialized knowledge involves examining the mathematics in the tasks that you select for your students and then using your mathematical and pedagogical content knowledge to predict the representational and computational strategies that they might use in responding to the tasks. Consider problem 1.1, for example:

> **Problem 1.1**
>
> Jamie solved the following problem for homework:
>
> > Three-fifths of the sand went through a sand timer in 18 minutes. If the rest of the sand goes through at the same rate, how long does it take all the sand to go through the sand timer?
>
> Jamie got 30 minutes for her answer. Show two different ways that Jamie could have solved the problem.

Reflect 1.1 asks you to examine problem 1.1 in the context of the Big Idea's focus on invariance. Later, Reflect 1.2 and 1.3 follow up on this work by asking you to analyze the work of three students with regard to the Big Idea.

Reflect 1.1

In what ways does invariance arise in problem 1.1?

How do you predict that students in grades 6–8 would attend to invariance in their solutions to the problem?

Problem 1.1 confronts students with the relationship between a quantity of sand (three-fifths of the sand) and the time for that quantity of sand to go through a given timer, assuming that the remainder of the sand will go through the timer at the same rate. That is, the ratio of "quantity of sand to time" for the remaining two-fifths of the sand to go through the timer is assumed to be equivalent to the ratio of "quantity of sand to time" for the initial three-fifths of the sand.

Underlying these two equivalent ratios is the tacit assumption that the ratio of *any* quantity of sand to time does not vary from the original ratio, *three-fifths to*

18 minutes. A further assumption is that for other timers to exhibit similar invariance in the ratios of quantity of sand in the timer to elapsed minutes for it to go through the timer, those other timers must be essentially identical to *our* timer.

If students approach the problem solely from the standpoint of the two equivalent ratios, they can reason that if $^3/_5$ of the sand takes 18 minutes to go through the timer, then $^2/_5$ of the sand must take 12 minutes to go through the same timer. They can determine that $^2/_5$ of the sand takes 12 minutes because they know that two-fifths is *two-thirds* of three-fifths, and *two-thirds* of 18 is 12. As figure 1.1 illustrates, by reasoning in this way, students assume only that the equivalence of the ratios is invariant: *three-fifths to 18* is equivalent to *two-fifths to the time for the remaining sand to go through the timer.*

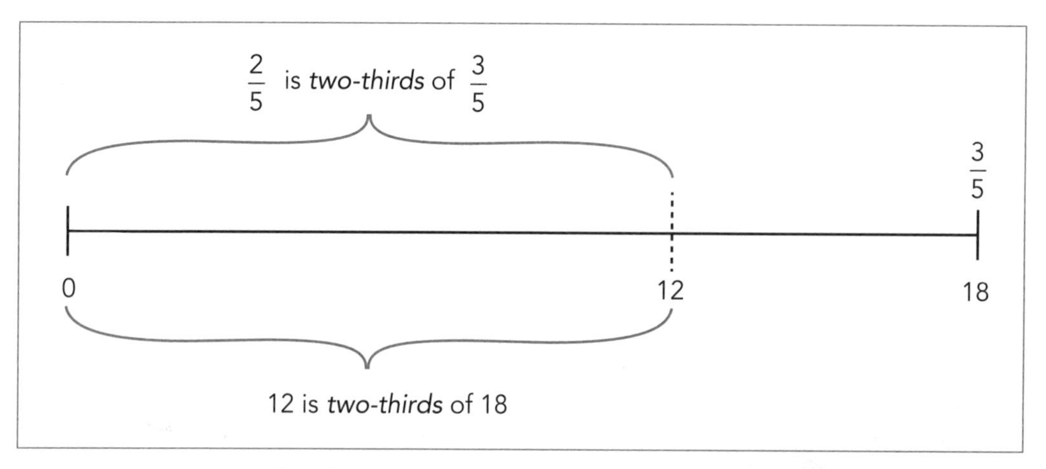

Fig. 1.1. Invariance in the relationship between the quantity of sand that has gone through a timer and the quantity of sand that remains to go through it, with respect to time required to go through the timer

The relationship between the original quantity of sand that went through the timer and the quantity of sand that remains to go through it should be the same as, or invariant from, the relationship between the original time taken and the remaining time needed. Namely, this invariant relationship is *two-thirds.* That is, the remaining quantity of sand, which is two-thirds of the quantity of sand that already went through the timer, should take two-thirds of the corresponding elapsed time. By reasoning about this relationship in this way, students can discern that the entire quantity of sand (five-fifths) should take the sum of the elapsed times for the respective amounts (three-fifths and two-fifths)—namely, 30 minutes—to go through the timer.

However, if students approach problem 1.1 by assuming that the first three-fifths of the sand went through the sand timer at a constant rate for any measurable amount of time, then the other invariant relationships will become evident. For example, given the ratio of $^3/_5$ of sand through the timer to 18 minutes, students may determine another ratio: $^1/_5$ of the sand through the timer to 6 minutes. Figure 1.2 illustrates one way in which students might determine the ratio $^1/_5$ to 6.

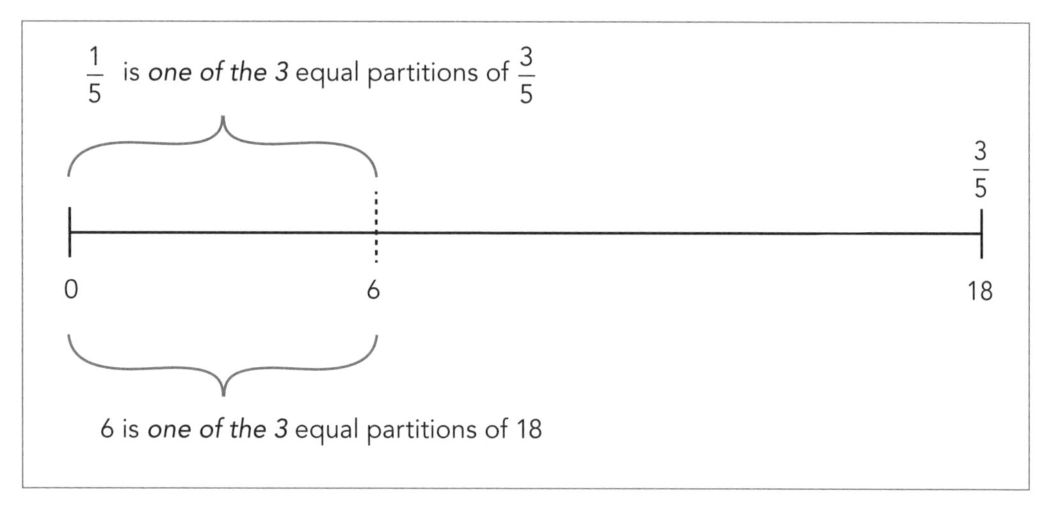

Fig. 1.2. Invariance in the relationship between a quantity of sand going through a timer and one part of an equal partitioning of that quantity of sand, with respect to elapsed time

Specifically, by identifying three equal partitions, each of $^1/_5$, students can determine the invariant relationship of *one to three*. Given that one of the partitions equaling $^1/_5$ is one-third the size of $^3/_5$, students can determine that one-third of 18 is 6. However, to understand the way in which Jamie may have gotten her answer of 30 minutes, they can now use this ratio to state that if one-fifth of the sand took 6 minutes to go through the timer, then five-fifths of the sand in the timer (the whole amount of sand in the timer) will take five times as many minutes—namely, 30 minutes (6 minutes taken 5 times, or 6 × 5).

The preceding discussions illustrate two of the possible ways in which Jamie might have interpreted the relationships in the problem. Were your predictions of how students might attend to invariance in explaining ways in which Jamie may have solved this problem similar to those in figures 1.1 and 1.2? If so, in what other ways could students attend to invariance? If not, how do your predictions match the students' work shown in figures 1.3–1.5? Use the questions in Reflect 1.2 to guide your examination of the work in these figures.

Reflect 1.2

Figures 1.3, 1.4, and 1.5 show work on problem 1.1 by three seventh-grade students, Marion, Judy, and Gottfried, respectively. How do these students attend to invariance in ratio relationships, as evidenced by their representations and explanations?

What do Marion, Judy, and Gottfried appear to understand about ratios and proportional relationships in the context of explaining Jamie's thinking?

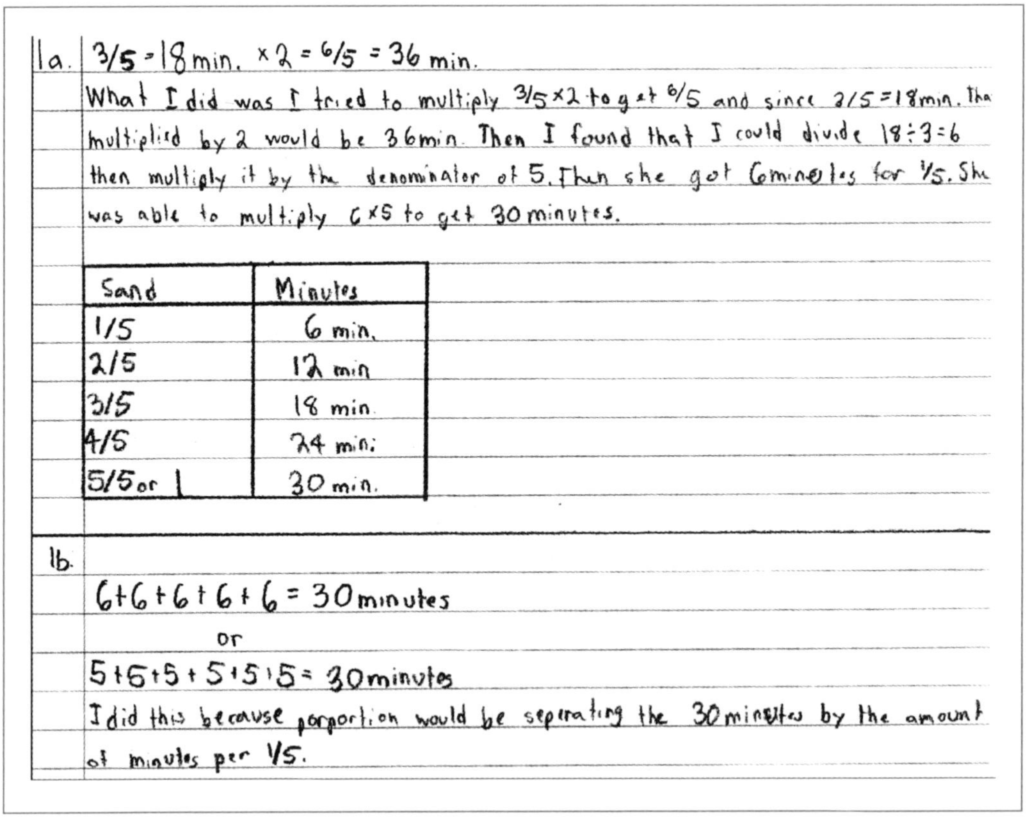

1a. 3/5 = 18 min. ×2 = 6/5 = 36 min.

What I did was I tried to multiply 3/5×2 to get 6/5 and since 3/5=18min. Tha multiplied by 2 would be 36min. Then I found that I could divide 18÷3=6 then multiply it by the denominator of 5. Then she got 6minutes for 1/5. She was able to multiply 6×5 to get 30 minutes.

Sand	Minutes
1/5	6 min.
2/5	12 min
3/5	18 min.
4/5	24 min.
5/5 or 1	30 min.

1b.

6+6+6+6+6 = 30 minutes

or

5+5+5+5+5+5 = 30minutes

I did this because porportion would be seperating the 30minutes by the amount of minutes per 1/5.

Fig. 1.3. Marion's response to problem 1.1

By examining students' work—other students' as well as your own—you can discover and explore new ideas and possible strategies for solving problems. To understand students' thinking, you must understand your own thinking—and many possible ways of thinking. Understanding multiple representations allows you to facilitate students' discussions of their own representations and strategies.

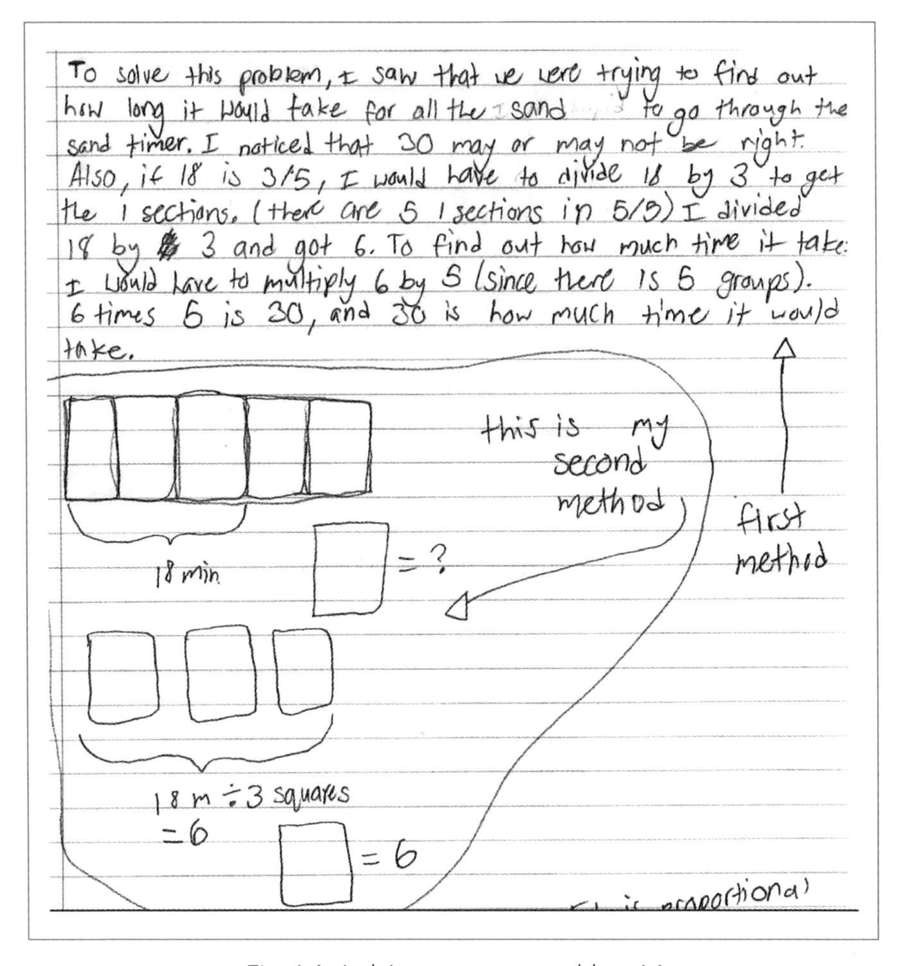

Fig. 1.4. Judy's response to problem 1.1

Figure 1.3 shows Marion's explanations of two ways in which Jamie might have solved problem 1.1. Notice that in the first way, Marion began by finding an equivalent ratio to $^3/_5$:18 by multiplying each part by 2. However, in the rest of her process, she did not rely on the new ratio, $^6/_5$:36. Instead, she divided 18 by 3 to get 6. In doing this operation, she recognized that she had identified a new ratio: $^1/_5$ of the sand through the timer to 6 minutes. That is, she partitioned 18 into three equal 6-minute sections. She then noted that to calculate 30 minutes, Jamie would have multiplied the 6 minutes by 5. In doing so, Marion recognized the invariance in the amount within each partition and the number of iterations required for all the sand to go through the timer.

One way that Jamie could have solved the problems is that 18 minutes is equal to ⅗. So what she could have done is that she could have divided 18 by 3. Then multiply that answer by 5 to get the time. The reason that I did this is that by dividing 18 by 3 will show how much ⅕ is equivalent to. Since there is 5 ⅕ inside of 5/5 then I multiplied the time that was equivalent to ⅕ by 5 and got 30.

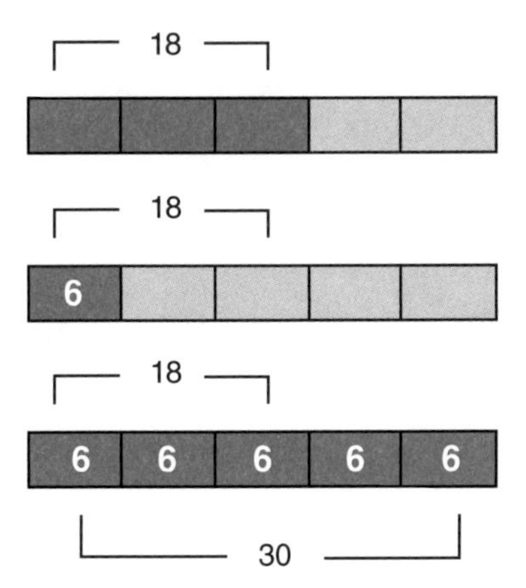

Another way to solve this problem is that you could solve the problem is that you could multiply 18 by 2. Then you could divide the number by 6. Then whatever the answer is, you would subtract that from 36 (18 x 2). The reason that I solved this problem this way is that because since 18 = ⅗ and when multiplying it by 2, you would get 6/5. Then since it is 6/5 it is one too much. So whatever number is equal to 6/5, you divide is by 6 since it is 6/5. Then when you get that number, you subtract that number from the 6/5 number and it will make it become 5/5.

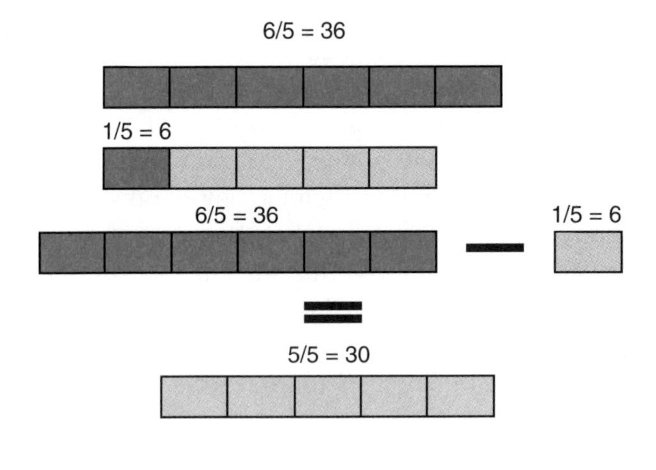

Fig. 1.5. Gottfried's response to problem 1.1

In showing a second way in which Jamie could have solved the problem, Marion appears to have suggested that Jamie might have taken an iterative approach to the problem. Although it is unclear how Marion determined that Jamie would iterate 6 five times, or 5 six times, her explanation seems to have focused on identifying quantities of minutes, iterations of which equal 30 minutes. However, the possibility does exist that Marion simply iterated 6 five times because she knew that this process yields 30.

Figure 1.4 shows Judy's ideas about two ways that Jamie might have solved problem 1.1. The approach that Judy used was similar to Marion's. Judy noted that 30 might or might not be the correct answer. In working to determine its correctness, she divided 18 minutes by 3. That is, like Marion, she partitioned 18 into three 6-minute sections. In her drawing, she labeled each of these sections as 6 minutes. However, in dividing 18 by 3, she also tacitly recognized that her partition of three-fifths of the sand would give her unit sections of sand (each equal to *one-fifth*), of which she would need 5 for all the sand to have passed through the timer. Much like Marion, Judy attended to invariance by partitioning one measure in the ratio (minutes) in the same manner as the other (quantity of sand).

The first solution that Gottfried offered in his work, shown in Figure 1.5, is similar to both Marion's and Judy's. In fact, in his explanation, Gottfried noted that he performed the division "to show how much $^1/_5$ is equivalent to." In his second method, he used the approach that Marion began with, when she multiplied each part in the ratio $^3/_5 : 18$ by 2. Unlike Marion, however, Gottfried pursued this idea and elaborated on his thought process. In particular, he argued that Jamie might have multiplied 18 by 2 to arrive at 36 minutes for $^6/_5$ of the sand to pass through the timer. Having obtained this new ratio of $^6/_5$ to 36, he then recognized the need to unitize the sand passing through the timer, and thus he divided by 6 to determine the ratio of $^1/_5$ of the sand going through the timer to 6 minutes. However, he eventually oriented himself to the same equivalent ratio of $^1/_5$ to 6 that he did in his first explanation of what Jamie could have done. That is, although Gottfried attempted to find out what was "one too much" with respect to $^6/_5$, he essentially partitioned $^6/_5$ into 6 equal sections, with each one equivalent to $^1/_5$, and similarly, he partitioned 36 into 6 equal sections of 6 minutes. He then subtracted one of these sections from 36 minutes to arrive at 30 minutes.

Throughout this process, Gottfried attended to invariance by iterating and partitioning. He first began by iterating the initial ratio, $^3/_5 : 18$, to arrive at $^6/_5 : 36$. He then partitioned 36 minutes in the same manner in which he partitioned $^6/_5$ to find the number of minutes needed for $^1/_5$ of the sand to pass through the timer. Although both Marion and Gottfried initially approached problem 1.1 through a process of doubling, Gottfried's representation seems to have allowed for more flexibility in conceptualizing the problem. Perhaps Marion's eventual use of a table limited her thinking largely to an additive perspective in relation to the unit.

The work of these three students illustrates their engagement in "unitizing" and "norming," processes that Lamon (1994) describes. Specifically, the students constructed a "reference unit," $1/5$, and "reinterpreted the situation in terms of that unit" (p. 92). In reconceptualizing the situation in terms of a ratio that is equivalent to the fixed $1/5$ unit, the students are also engaging in the norming process. Lamon describes the method that they employ in this norming process as the "within" strategy for interpreting ratio relationships. In particular, Lamon (1994, p. 95) notes that a student is "using a within strategy or scalar method when he or she equates two internal or within-measure space ratios and uses the sameness of scalar operator to determine the missing term." Lobato and Ellis (2010) present this method as proportional reasoning method 1 (pp. 10–11), which, in the case of problem 1.1, could involve the following steps:

- Forming a ratio by thinking of $3/5$ of the sand as 3 of the 5 parts of sand passing through the timer in 18 minutes.
- Partitioning the 3 parts (dividing by 3) to get $1/5$ (1 part) of the sand passing through in 6 minutes.
- Iterating (building up) units of $1/5$ of the sand and 6 minutes of time to get the whole amount of sand passing through in 30 minutes.

In the work shown in figures 1.3–1.5, the initial interpretation that the students made focused on the relationship *within* the measure spaces associated with $3/5$ and 18. In particular, Judy and Gottfried tacitly linked $3/5$ of the sand and 18 minutes, and at some point in their solutions they recognized that 18 minutes divided into 3 equal parts would give 3 intervals, each 6 minutes long. They then recognized that to calculate five of those intervals, they could use additive reasoning to determine the equivalent ratio $5/5:30$. Hence, Marion, Judy, and Gottfried all appear to have approached reasoning proportionally from a *within* perspective because they tried to understand what was happening with only one of the units—in this case, time—and then expressed the same *invariant* relationship in the other unit—in this case, amount of sand. As Lobato and Ellis might note in such a case, the students partitioned 18 minutes into three 6-minute sections and then iterated that partition five times to account for the passage of all the sand through the timer.

Gottfried's first response suggests "within" reasoning from a partitioning perspective. In his second response, in which he used the process of doubling the values of the numbers in the ratio $3/5$ and 18 to arrive at the ratio $6/5:36$, Gottfried also employed "within" reasoning from an iteration perspective. That is, Gottfried was looking "within" to identify a ratio close to $^5/_5$ by doubling $^3/_5$ and therefore also doubling 18. Once he arrived at the ratio $^6/_5$:36, he recognized that he had "one too much" and subtracted "$^1/_5$ = 6," an equivalence that he most likely observed in his original partitioning solution strategy.

Lamon (1994) and Lobato and Ellis (2010) identify another method of reasoning about proportional relationships. As Lamon notes, "A student is using a *between strategy* or a *functional method* when he or she equates two external or between-measure space ratios and relies on the functional relationship between the measure spaces to find the missing term" (p. 96). Lobato and Ellis (2010) identify such a strategy as proportional reasoning method 2 (p. 11). Applying this method to problem 1.1 could involve the following steps:

- Comparing the two numerical values $^3/_5$ and 18 (from $^3/_5$ of the sand in 18 minutes) by finding out how many times greater 18 is than $^3/_5$. Eighteen is 30 times greater than $^3/_5$. This gives an invariant functional relationship between the numbers associated with the amount of sand and the amount of time.
- Determining the amount of time for any quantity of sand to pass through the timer by multiplying the value of the quantity by 30. For example, $^1/_5$ of the sand would pass through the timer in $^1/_5 \times 30$, or 6, minutes, $^1/_3$ of the sand would pass through the timer in $^1/_3 \times 30$, or 10, minutes, $^2/_5$ of the sand would pass through the timer in $^2/_5 \times 30$, or 12, minutes, ..., $^5/_5$ of the sand would pass through the timer in $^5/_5 \times 30$, or 30, minutes, ...

With this approach to problem 1.1, students would directly compare quantities of sand with quantities of time. They would find and use the invariance of the 1:30 ratio either to solve for the time that it would take any amount of sand to pass through the timer or to solve for the amount of sand that would pass through the timer in any given amount of time. However, analysis of students' responses to this problem showed that no students used this strategy. It is possible that the context or wording of the problem made a "within" strategy a more natural way for students to develop a solution strategy.

When students use "between" reasoning to approach this problem, they identify the invariant relationship *between* the measure spaces. One question that they might ask is, "How many times greater is 18 than $^3/_5$?" Alternatively, they might ask, "How many times smaller is $^3/_5$ than 18?" By maintaining the factor of 30, students can construct a set of equivalent ratios for which they can determine the time for any given quantity of sand left in the timer to pass through it. In this manner, as previously discussed, the time needed for the quantity of $^5/_5$ of the sand to pass through the timer would be $^5/_5 \times 30$, or 30 minutes.

It is worth noting that Marion organized her thinking in a table format (sand and minutes)—a form often associated with determining a functional relationship. When a student creates such a table representation, the teacher may be tempted to attribute functional reasoning to his or her thinking process. However, at no time in her explanation did Marion examine relationships *between* the measure spaces (sand and minutes) to identify the scale factor of 30. Rather, Marion most likely organized her iterations of $^1/_5$ and 6 in the table to keep track of the possible equivalent ratios for each one-fifth of the sand through the timer, or for each 6 minutes.

Summarizing Pedagogical Content Knowledge to Support the Big Idea

Teaching the mathematical ideas in this chapter requires a specialized knowledge related to the four components presented in the Introduction: learners, curriculum, instructional strategies, and assessment. The four sections that follow summarize some examples of these specialized knowledge bases in relation to the Big Idea of ratios and proportions: invariance. Although we separate them to highlight their importance, we also recognize that they are connected and support one another.

Knowledge of learners

Lamon (1993) summarizes the work of Lesh, Post, and Behr (1988) as she reports on sixth-grade students' thinking related to ratio and proportion content. In the process, she provides a clear statement of the importance of this mathematical domain: "Proportional reasoning plays such a critical role in a student's mathematical development that it has been called a watershed concept, a cornerstone of higher mathematics, and the capstone of elementary concepts" (p. 41). She cautions, however, "It has been implicitly understood that proportional reasoning consists of being able to construct and algebraically solve proportions" (p. 41).

As a teacher, you must recognize that your students may arrive in your classroom with the expectation that working with and reasoning about ratios and proportions essentially entail constructing proportions symbolically and solving them algebraically. Of many available constructs, Lamon (1994) provides two that can help teachers understand the strategies that students use in tasks that require them to focus on and account for invariance in working with equivalent ratios and engaging in proportional reasoning.

Specifically, Lamon (1994) identifies *unitizing* and *norming*—two processes that students use to develop their mathematical thinking. Unitizing is students' use of a unit to reorient quantities within the mathematical context of the situation. Although this activity "probably begins visually in early childhood quantifying activities" (p. 92), working with multiplicative structures requires coordination of mental constructs. As a result, the activity of unitizing in proportional reasoning is not simple for students in the middle grades. Lamon (p. 93) considers the "simple" case of finding $^3/_4$ of 16 objects. She notes that to coordinate their thinking, students must reunitize at least three times:

1. Consider the 16 objects as 16 units. They then have 16 one-units.
2. Create units of units—that is, 4 composite units, each consisting of 4 one-units. They then have 4 four-units.
3. Create units of units of units—that is, create one three-unit consisting of three of the four four-units.

The samples of student work in this chapter highlight this reunitizing process. In the work shown for problem 1.1, the students recognized three-fifths, or 18, as one unit. They then created units of units—that is, three units, each consisting of either $^1/_5$ or 6, depending on the way in which they approached the problem. Finally, they created units of units of units when they created one $^5/_5$-unit consisting of five of the $^1/_5$-units, or one 30-unit consisting of five of the 6-units. As a teacher, you need to understand that the students approached their *unitizing* processes through understanding the relationships *within* measure spaces. However, this chapter has also suggested how students could examine the problem by examining functional relationships *between* measure spaces.

In planning opportunities for your students to navigate invariance in ratios and proportions, bear in mind a fundamental truth: each student's understanding of these concepts will be different from every other student's, and, perhaps even more important, different from your own understandings. Steffe's (1994, p. 5) observation underscores and extends this point:

> Hence, there is no reason whatever to assume that the child will interpret the formulation of a given problem in the way that seems "obvious" to the adult or the teacher; nor is there any reason to assume that the child could "see" that a particular way of proceeding must lead to a result that constitutes the solution.

Knowledge of curriculum

Selecting tasks that will allow your students to work from their own understanding to develop solution strategies that can engage their peers in robust discussions is an integral aspect of teaching that draws on your knowledge of curriculum. Such tasks do not typically have a prominent place in the classroom, and they will not unless you recognize the need to challenge students with them and allow time for discussion of them. A comment by Steffe (1994) is arguably as pertinent today as it was twenty years ago, if not more so: "The current notion of school mathematics is based almost exclusively on formal mathematical procedures and concepts that, of their nature, are very remote from the conceptual world of the children who are to learn them" (pp. 5–6). As a teacher, you must ask yourself, "What is the conceptual world of students who are to reason about and solve problems involving ratios and proportions in grades 6–8?"

Your answers to this question can focus on students' general strategies, as discussed in this chapter, using "within" and "between" strategies. You must also consider the conceptual world of students with respect to invariance, as well as fraction concepts as they relate to ratios. *Developing Essential Understanding of Rational Numbers for Teaching Mathematics in Grades 3–5* (Barnett-Clark et al. 2010) and the companion volume, *Putting Essential Understanding of Fractions into Practice in Grades 3–5* (Chval, Lannin, and Jones 2013) outline connections to the conceptual world of middle school students with regard to rational numbers and fractions. The ways in which students have developed prior understandings of partitioning and composing units will critically affect the conceptual worlds that they inhabit when they enter your classroom.

More important, those conceptual worlds will almost certainly be different from your own, developed over years of learning, using, and teaching these concepts. Consequently, it is critical to encourage students to demonstrate more than one way in which they can solve a problem. Seeing students' multiple strategies will give you access to the robust ways in which students are approaching problems and allow you to determine whether they are able to work a problem from various perspectives, such as "within," "between," additively, and multiplicatively. Knowing where students' abilities lie is a critical component in providing focus for future mathematical work.

In the companion volume to this one, *Developing Essential Understanding of Ratios, Proportions, and Proportional Reasoning in Grades 6–8*, Lobato and Ellis (2010) offer a thoughtful analysis of the essential understandings related to invariance. They provide two methods for engaging in proportional reasoning that align with Lamon's "within" and "between" strategies. Furthermore, they present tasks throughout the book that highlight invariance in proportional reasoning from a variety of perspectives.

This chapter and those that follow offer tasks that build on the examples and analysis provided by Lobato and Ellis and challenge you to think about tasks that will provide your students with varied and rich opportunities to learn proportional reasoning concepts. In this chapter, we have presented a problem situation that we believe highlights concepts of invariance while allowing for multiple interpretations of Jamie's solution strategies. As students consider how Jamie might have solved the problem, they are encouraged to provide multiple solutions. When you ask your students to show their strategies, you assume responsibility for recognizing why and how their strategies differ. If a student presents two solutions, are they different because of the computations performed? (For example, consider multiplying $^1/_5$ five times to arrive at $^5/_5$ and multiplying 6 five times to arrive at 30.) Has the student presented two fundamentally different approaches to proportional reasoning? (Consider, for example, unitizing 6 to form three equal partitions of 18 because making three equal partitions is also a useful way to partition $^3/_5$ into one-fifth portions, or unitizing $^5/_5$ because it is five-thirds as large as $^3/_5$, or recognizing that $^2/_5$ is two-thirds as large as $^3/_5$—all "within" strategies—or recognizing the constant scale factor of 30 between the numbers involved in different measures, because 18 is 30 times as large as $^3/_5$, and using it to identify a set of equivalent ratios, one of which is $^5/_5$: 30—a "between" or "functional" strategy.)

In selecting tasks, you should be careful to identify opportunities that will not perpetuate students' misconceptions. Giving tasks that engage your students in thinking about the multiple ways in which ratios are equivalent can lead them to develop misconceptions *unless* you help them give careful attention to the various invariant relationships in the context. Suppose that you present the following task:

Punching Up the Recipe

Lee is having a party for many of his friends. He is making punch from a favorite recipe. However, with so many friends coming to the party, Lee realizes that the recipe will make only $^1/_4$ of the full amount he will need. The recipe calls for 3 cups of orange juice and 4 cups of pineapple juice. Lee makes the full amount that he will need. After tasting the completely mixed batch of punch, Lee decides that it doesn't have enough pineapple taste. To get the taste that he wants, he adds 4 cups of pineapple juice. He likes the taste of the resulting punch and decides to modify his recipe for punch. What is the ratio of orange juice to pineapple juice in Lee's new recipe?

In addition to potentially promoting a diversity of student thinking and strategies (both computational strategies and *norming* strategies), the task draws attention to two invariant relationships. One invariant relationship is the ratio 1:4 between the recipe and the full quantity of punch needed for the party, and the other invariant relationship is the ratio 3:4 between orange juice and pineapple juice in the punch made according to the original recipe. However, in this problem, students are asked to reunitize the full quantity of punch needed for the party and then include 4 more cups of pineapple juice than called for in the original recipe. A new invariant relationship is formed in the ratio of orange juice to pineapple juice in the full quantity of punch needed. This problem has the potential to reveal misconceptions that students have about how the additional 4 cups of pineapple juice affect the ratio. Some students might give 3 : 8 as the ratio in the new recipe, erroneously reasoning that the original recipe called for orange and pineapple juice in a ratio of 3 to 4, and then 4 more cups of pineapple juice were added, making the ratio 3 to 8. By contrast, students might solve this problem correctly in several ways. For example, they might calculate 4 times the amounts of each ingredient in Lee's original recipe to determine that the quantity of punch needed for the party contained 12 cups of orange juice and 16 cups of pineapple juice, and then they could add 4 cups of pineapple for a new ratio—12 : 20.

A classroom discussion could be productive at this point, serving to identify students who gave the answer as 12 and 20 and students who gave it as an equivalent ratio such as 3 and 5. Such a discussion would allow students to exchange ideas about their reasoning, showing one another how they were reasoning about ratios. It could also reveal which students provided their answer in terms of cups, and which were thinking in terms of equivalent ratios.

With any task, this one included, thinking about the different possibilities for solving the task is important. This means inviting yourself, as the teacher, to work on the task from as many perspectives as you are able. For example, with the punch problem, how might you think of it from a quadrupling perspective—that is, to think about the given recipe as being quadrupled? In this manner, if 4 cups of pineapple juice are added to the quadrupled amount, can you reason that 1 cup would be added to the original recipe, giving 3 cups of orange and 5 cups of pineapple?

Attending to students' thinking when they are engaging in discussions of invariance is important. Often different invariant relationships exist, but students do not attend purposefully to them. Instead, they identify only the ratio that is written symbolically with a colon. In these cases, you must offer guidance, either through integrating other students' strategies into the discussion or by employing other instructional strategies. Whether students are answering only in one way or providing a variety of solutions, you need appropriate instructional strategies to highlight, challenge, and build individual students' understandings meaningfully through discussions and classroom interactions—of your students with one another and with you.

Knowledge of instructional strategies

This chapter has provided samples of instructional strategies that you can use to develop your students' mathematical understandings by focusing on their thinking and strategies. In particular, including tasks that, like problem 1.1, encourage students to explain another student's thinking is a strategy that potentially leads to other instructional decisions. If, in response to this problem, students in your class provide only explanations that are similar to those of Marion, Judy, and Gottfried, then you should decide how to bring other possible solutions into a general discussion. Lamberg (2012) offers important strategies for facilitating whole-class discussions. Through engaging your students in "math talk" in peer, small-group, and whole-class discussions, you enable them to work collaboratively to devise other strategies that you can highlight as you facilitate the conversations.

Lamberg (2012) notes particular characteristics of discussions that facilitate the development of students' understanding:

- Students develop "shared" understanding of a problem.
- Teachers provide students with questions that support students' metacognitive processes.
- Students evaluate and analyze their thinking and their peers' thinking.

Furthermore, she advocates (p. 11) that teachers go through three phases in the course of facilitating mathematical connections in whole-class discussions:

Phase 1: Making thinking explicit

Phase 2: Analyzing solutions

Phase 3: Developing new mathematical insights.

In all phases, the teacher facilitates the discussion through questioning.

Similarly, Smith and Stein (2011) identify five practices for productive discussions. Specifically, they elaborate on the ways in which teachers can orchestrate mathematical discussions through—

- anticipating students' solutions;
- monitoring students' in-class work;
- selecting approaches for students to share;
- sequencing the approaches purposefully to maximize impact and learning; and
- connecting the approaches and the underlying mathematics.

By making your students' thinking explicit and connecting the ways in which they are partitioning, composing, and working with units when dealing with ratios, proportions, and proportional reasoning, you highlight for them ways in which the relationships are invariant *within* and *between* measure spaces (for instance, the amount of sand, or the minutes elapsed in problem 1.1). However, Lamberg's (2012) three phases and Smith and Stein's (2011) five practices involve knowledge of assessment and related student understandings, along with what Lamberg (p. 11) identifies as "planning during the lesson (Assess student reasoning/errors/misconceptions. Identify topic for discussion/problem to discuss and thinking about how to sequence discussion)."

Knowledge of assessment

Wiliam (2007) highlights the importance of assessments of the formative type—that is, assessments that allow students to share their thinking and teachers to make prompt instructional decisions based on the understanding and misunderstanding that the students demonstrate. As this chapter has illustrated in the samples of work from Marion, Judy, and Gottfried, students' work can provide insight into their approaches to invariance in their proportional reasoning. Such insight is possible only when teachers ask students to show their thinking on a task that encourages reversibility, flexibility, and generalization in students' thinking (Dougherty 2001), as discussed in the Introduction.

Considering assessment possibilities in determining tasks to give students is critical. Identifying avenues for follow-up questions that allow for further assessment of student thinking is essential. Furthermore, the purpose of prompting students to give multiple solution strategies is not only to show them that problems have a variety of potential solution paths but also to enable you to conduct a critical aspect of assessment. In providing multiple strategies, students display the diversity of their own thinking or reveal the possible limitations of the connections that they are making. In other words, as a teacher, you must push your students to provide more than one solution path to give you a better understanding of their ability to use and develop appropriate strategies and tools from more than one solution perspective. When students are reasoning in ways that are similiar to the three students whose work is highlighted in this chapter, a teacher with knowledge of the mechanisms of unitizing, norming, partitioning, composing, and the "within" and "between" strategies can ask follow-up questions to clarify their thinking. As discussed in the previous section, such follow-up questions can be directed to students individually or in peer, small-group, or whole-class discussions. If your students are examining a problem from one perspective, you should allow them an opportunity to rethink or reformulate their strategies, prompting them with questions such as, "Can you talk to me or your partner(s) about other relationships you see to $^3/_5$ (or 18) in this problem?"

Conclusion

The figures and students' work in this chapter have illustrated mathematics that teachers should grasp to understand their students as individual learners, design learning opportunities for them, make critical instructional decisions, and understand their responses to assessments in meaningful ways. As a teacher, you need to understand ways in which your students are currently attending, and may potentially attend, to invariance in relationships when learning ratios and proportions and reasoning about them.

The chapters that follow highlight essential understandings from *Developing Essential Understanding of Ratios, Proportions, and Proportional Reasoning for Teaching Mathematics in Grades 6–8* (Lobato and Ellis 2010) that are particularly challenging to put into practice. In all these discussions, although invariance is not always specifically addressed, this overarching Big Idea is ever present.

into practice

Chapter 2
Using Ratios to Measure Attributes

Essential Understanding 3
Forming a ratio as a measure of a real-world attribute involves isolating that attribute from other attributes and understanding the effect of changing each quantity on the attribute of interest.

Measuring an attribute that involves a relationship between quantities often requires dealing with real-world or physical contexts. To work with ratios as measures of attributes, students must understand such attributes conceptually and be able to recognize how their measures change as each quantity in a ratio changes, as Essential Understanding 3 suggests.

This chapter uses the term *attribute* to describe a measurable feature of an object or situation. Attributes such as the length of a soccer field or the mass of a person are such characteristics (Michell 2005) and can be measured directly. These measures are *additive*, meaning that two measured parts of the attribute can be added together to make an increased measured amount of the attribute. Three cups of milk can be added to five cups of milk to make eight cups, or a half-gallon, of milk.

Other measurable attributes are *relationships* (Michell 2005), such as speed, which is an attribute of the relationship between distance and time with respect to a traveling object. A car that travels steadily for four miles in five minutes is traveling 0.8 miles every minute. The appropriate measures for attributes that are relationships are *ratios* (Sowder et al. 1998); one quantity of the attribute is always a certain number of times the other quantity (Michell 2005).

The attributes examined in this chapter are set in contexts. Contextualizing otherwise "bare" mathematical problems in situations from the everyday world helps students relate to the mathematical ideas in the problems. Situating mathematics problems in contexts, however, can have a confounding effect on a student's solution process (Lo and Watanabe 1997). If the context is familiar, students may be distracted by previous experiences involving that situation. For example, students may reasonably think that running up a road with a 4 percent grade for 0.5 miles is harder than running up a road with a 4 percent grade for 0.25 miles. However, this reasoning is not helpful when solving certain mathematics problems, such as one that involves comparing the grades of the two roads.

In fact, when high school students were asked to measure the steepness of wheelchair ramps (Lobato and Thanheiser 2002), the majority of them considered a longer ramp to be steeper than a shorter ramp of the same steepness because they believed that wheeling up the greater distance would be more tiring (see Lobato and Ellis 2010). Sowder and her colleagues (1998) suggest that this type of thinking indicates an inability to mathematize a situation, a process that requires students to "understand the relationship between the mathematical formulation—in this case, ratio—and the physical situation that it represents" (Sowder et al. 1998, p. 140). The students in Lobato and Thanheiser's study could not formulate a mathematical model involving a ratio of the height of the ramp to the length of the base, which was related to the attribute of steepness for the ramp.

This chapter's discussion of work with attributes that are measured as ratios focuses on sixth- and seventh-grade students' work on three tasks. The first task, presented as problem 2.1, is designed to prompt sixth-grade students to develop an awareness of attributes that are properties that are measured directly—in this case, the length and area of objects. Students reveal their assumptions about these properties by considering the multiple attributes that one could use to describe the relative sizes of two polygons in the problem.

The second and third tasks discussed in the chapter involve working with measurable attributes that are relationships. Conceptual understanding of attributes that are measured as relationships, or ratios of one attribute to another, develops later than understanding of attributes that are measured directly, on their own (Nunes, Desli, and Bell 2003). The tasks presented as problems 2.2 and 2.3 are thus more challenging for middle-grades students than problem 2.1. Problem 2.2, given to seventh graders, asks students to determine from which of two bags one would be more likely to draw a gumball of a particular color. This chapter's discussion invites examination of students' work to discern the ways in which their rationales offer insights into their conceptualizations of the relevant quantities.

The last task, problem 2.3, also given to seventh-grade students, calls for analyzing a situation to determine which of two pizza chefs would win a pizza-making speed contest. Again, the discussion examines students' work, revealing that students frequently agreed on the faster chef, but their approaches to the problem varied, highlighting different solution strategies and ways of thinking.

Working toward Essential Understanding 3

How can you help your students develop the quantitative reasoning needed to consider attributes that are measured as relationships, such as the sweetness of a pitcher of tea? How can you support them in understanding that the sweetness is affected by a change in the volume of water or amount of sugar? What happens to the sweetness of the tea if someone adds more water, but more of nothing else, to the pitcher? To explore ways to help students think about attributes that are relationships, this chapter begins with a simpler problem.

Using direct measurements of an attribute to compare objects

In problem 2.1, students were asked to decide which of two shapes is "larger" and explain the reasoning behind their choice:

Problem 2.1

Molly and Polly were looking at the two figures pictured below. They were trying to decide which figure is larger.

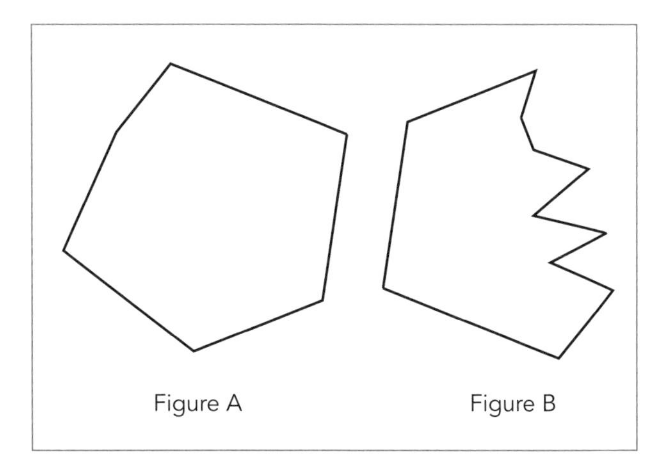

Which figure is larger? Explain your thinking.

This problem was designed to prompt students to discuss the attribute or attributes that they focused on in deciding which of two polygons is larger. What assumptions did they make, and what were their conceptions of how the attribute relates to the size of the polygon?

Sixth-grade students solved this problem at the start of a measurement unit in which they were studying area and perimeter. The task was designed specifically to focus the students' attention on attributes that they might use in comparing the polygons with respect to size. The word "larger" was deliberately chosen for its ambiguity, to force students to take a particular perspective and explain their solution process from that point of view. Further, the design of the polygons was intended to encourage students to invent new methods for quantifying or measuring the attribute or attributes that they chose to use to make the comparison. This type of open-ended problem is likely to lead to a class discussion about the assumptions that students make—in this case, as they decided on the larger polygon. A problem of this sort also enables the teacher to assess the prior experience and understanding that students bring to this topic.

Figures 2.1–2.3 show three sixth-grade students' work on problem 2.1. Use the questions in Reflect 2.1 to guide your examination of their strategies and solutions.

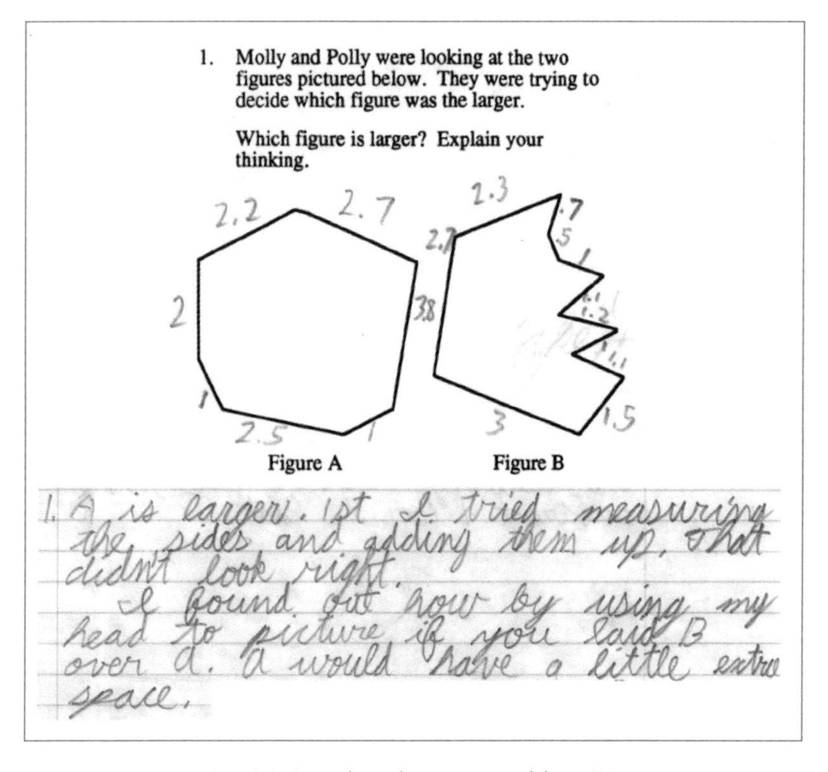

Fig. 2.1. David's solution to problem 2.1

Reflect 2.1

Work on problem 2.1 by three sixth graders—David, Thomas, and Evangeline—appears in figures 2.1, 2.2, and 2.3, respectively. What can you glean about the students' conceptualizing of the attributes from their responses to the task?

What assumptions did each student make in deciding which polygon is larger?

What questions might you ask in a class discussion to address the differences in the students' approaches and comparisons?

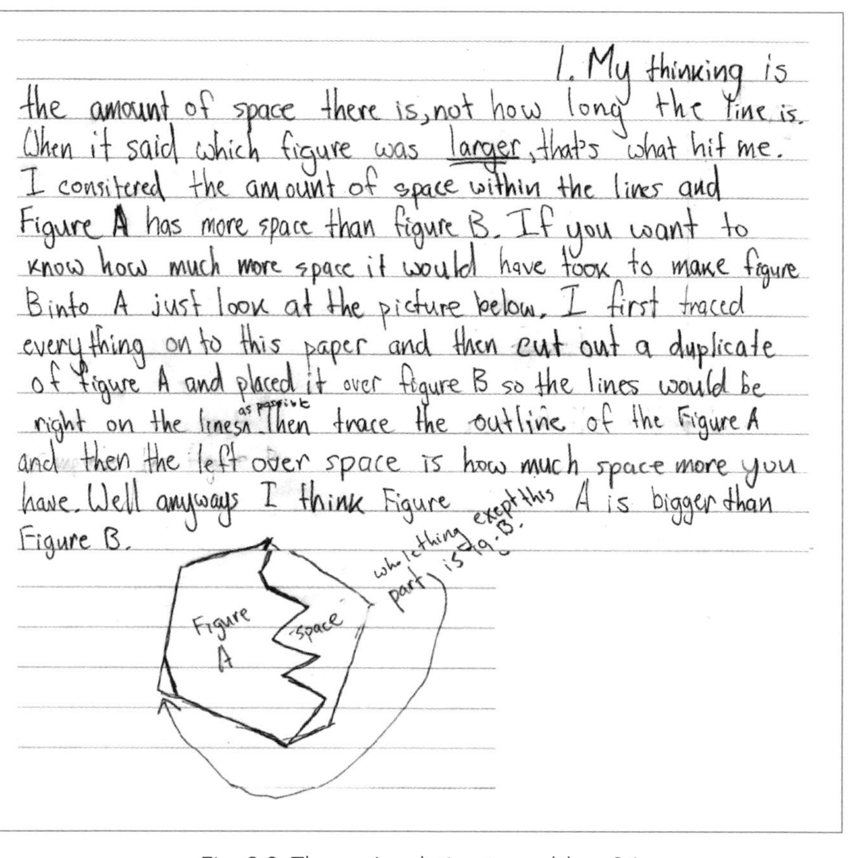
1. My thinking is the amount of space there is, not how long the line is. When it said which figure was larger, that's what hit me. I consitered the amount of space within the lines and Figure A has more space than figure B. If you want to know how much more space it would have took to make figure B into A just look at the picture below. I first traced everything on to this paper and then cut out a duplicate of figure A and placed it over figure B so the lines would be right on the lines as possible. Then trace the outline of the Figure A and then the left over space is how much space more you have. Well anyways I think Figure A is bigger than Figure B.

Fig. 2.2. Thomas's solution to problem 2.1

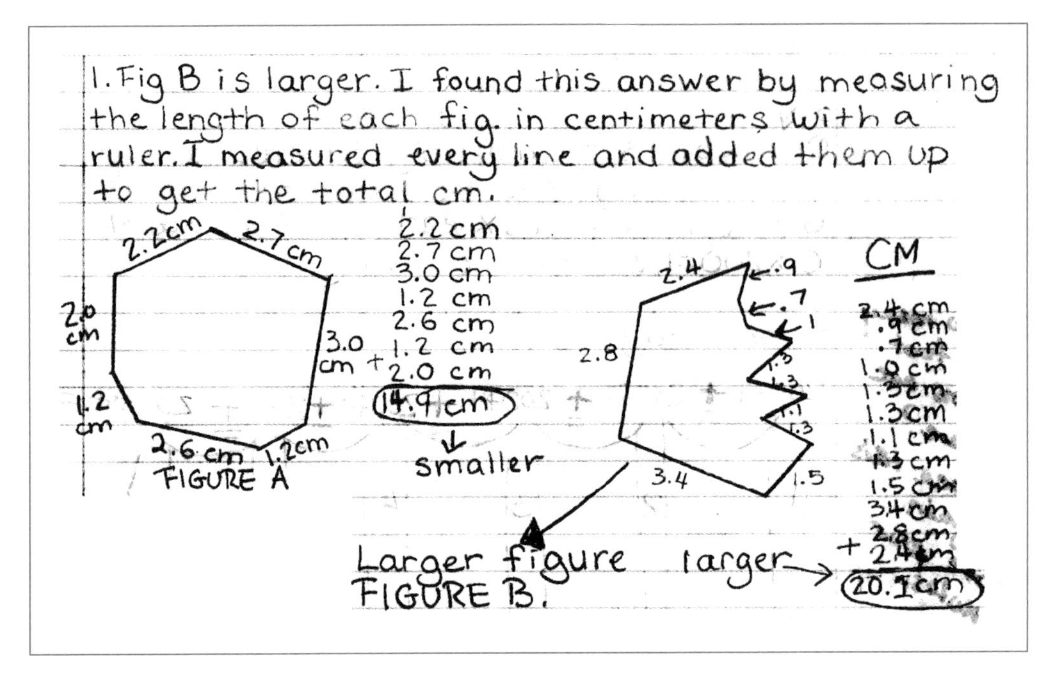

Fig. 2.3. Evangeline's solution to problem 2.1

David's response, shown in figure 2.1, indicates that after trying one mathematical approach to the problem, he abandoned it and then made a judgment that figure A is larger on a different basis (after "using [his] head"). His original approach was to measure the line segments in the two polygons, but he was troubled when the measure of the perimeter of figure B turned out to be greater than the measure of the perimeter of figure A. Clearly, David initially intended to base his determination on the attribute of perimeter. However, what appeared to him to be the greater area of figure A seems to have been a compelling factor in David's assumption that "larger" in the question means covering more surface. Many would probably make a similar assumption in this situation. David writes in his explanation, "A is larger. 1st I tried measuring the sides and adding them up. That didn't look right. I found out how by using my head to picture if you laid B over A. A would have a little extra space." The disequilibrium that David experienced when his solution "didn't look right" did not cause him to reflect on his initial assumption. He did not consider that figure B's greater perimeter could be relevant to the question. Instead, he found a strategy that would give him the result that he was expecting—namely, that figure A is "larger" because it has more area than figure B.

Thomas, whose work appears in figure 2.2, also made an assumption about which attribute of the polygons one should compare to determine the larger one. The first sentence of his explanation suggests that he recognized that someone might consider determining "how long the line is" as a way to decide which polygon is larger, but he rejected that possibility and did not consider it valid: "My thinking is the amount of space there is, not how long the line is." So he chose a different approach. Thomas's strategy for comparing the areas was to "cut out a duplicate of figure A and place it over figure B," matching the perimeters where possible. His diagram shows the difference in the areas of the two polygons, with his small note next to the drawing indicating that the "whole thing"–that is, all of figure B except for one small corner part–overlapped figure A. Thomas's explanation shows that his understanding of the attribute–in this case, area–did not rely on a procedure for calculating the numerical value. Although he did not mention "area" but referred instead to "space," the method he used to compare the polygons makes it clear that he had an understanding of the concept of area. Thomas focused on the attribute of area in his comparison of the two polygons without using a formula or other means of counting.

Evangeline, whose work is shown in figure 2.3, solved the problem through an approach that is similar to David's initial one. She explained, "Figure B is larger. I found this answer by measuring the length of each fig. in centimeters with a ruler. I measured every line and added them up to get the total cm." Unlike David, however, she was not perplexed or confused by her findings. She simply assumed that figure B is larger than figure A because the total distance around it, as measured in centimeters, is greater. During the subsequent class discussion, one of the other students asked Evangeline how she came up with the idea of using "the outside line" to decide which polygon is larger. She replied, "I thought of myself as a tiny ant who had to walk along the outside. It would take me a lot longer to walk around figure B." Evangeline contextualized the attribute of perimeter kinesthetically, in the notion of taking a walk on the boundary lines of each of the polygons. Comparing the sizes of the polygons to determine which is larger was, for her, a matter of comparing how long it would take to traverse the perimeters of the polygons–not a matter of comparing the "space" (area) in the polygons–the notion that was foundational to David's and Thomas's conceptualizations of the problem.

Evangeline took some notes during the class discussion of the problem. Her notes, shown in figure 2.4, indicate the multiple ways in which students in this class interpreted the problem and approached its solution. Students came to different conclusions: the polygons are the same size (determined by considering height), figure B is larger (determined by considering perimeter), and figure A is larger (determined by considering area).

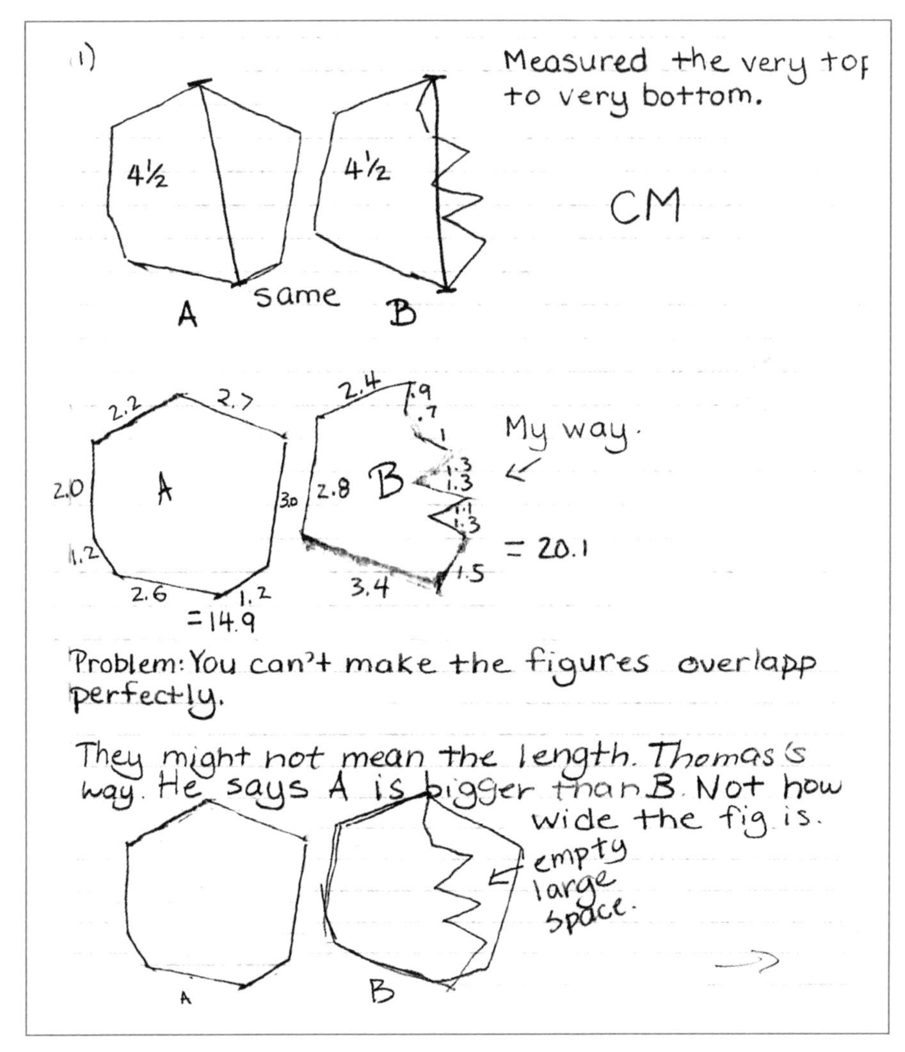

Fig. 2.4. Evangeline's notes from the class discussion of students' work on problem 2.1

The teacher extended the discussion by posing a series of questions to focus students' attention on different attributes that they could have considered in determining the larger polygon:

- "What did the word *larger* mean to you?"
- "What were you looking at to make your decision about which figure was larger?"
- "Once you decided what to compare, how did you make the comparison?"

These questions addressed three important aspects of comparing the polygons: making assumptions about the situation, determining the relevant attributes, and quantifying or comparing that attribute with an appropriate measurement method. The note that Evangeline made next to her drawing representing Thomas's method (see fig. 2.4) expresses a significant realization: "They [the problem's authors] may not mean length." This statement indicates a critical awareness that when comparing mathematical objects, being purposeful about what attributes to consider and why is foundational to recognizing important mathematical relationships that can be explored.

Moving to relational attributes: Probability and speed

Relational attributes present students with new challenges. Problem 2.2 confronts them with a situation in which the attribute of interest is likelihood, or probability:

> **Problem 2.2**
>
> Taufique has two bags, each containing some red and some white gumballs. In bag A, there are 10 red and 15 white gumballs. In bag B, there are 6 red and 8 white gumballs.
>
> If Taufique reaches in without looking, from which bag is he more likely to pull out a red gumball?
>
> Draw a diagram to represent how you solved the problem, and explain it.

The attribute of probability is a relationship: the likelihood of a particular outcome occurring (in this problem, pulling out a red gumball) is the ratio of the quantity of a particular possibility to the quantity of all the possibilities (Bryant and Nunes 2012). We may use different techniques for calculating a probability, but our method must be based on all possible outcomes and not just the number of ways that the event we want to predict can occur (Bryant and Nunes 2012).

Studies (Nunes, Desli, and Bell 2003; Bryant and Nunes 2012) show that young students often base their analysis of probability situations on one quantity only—usually the quantity involving the outcome of interest. Before turning to the ways in which seventh-grade students worked with the quantities of red and white gumballs in problem 2.2 to predict the probability of drawing a red gumball from a bag, consider the question in Reflect 2.2.

Reflect 2.2

How do you suppose the numbers of red and white gumballs in bag A and bag B might affect students' thinking about the situation?

The greater number of red gumballs in bag A often influences young students' responses (Bryant and Nunes 2012). However, in the group of seventh-grade students whose work we examined, the majority of those who solved this problem were not distracted by the number of red gumballs in bag A as compared with the number in bag B. The students whose solutions we include were able to identify bag B correctly as the one that would offer someone the greater probability of pulling out a red gumball. However, the students' explanations of why they chose bag B reveal variations in their mathematical reasoning. It is important to note that these students completed this problem at a point in their study of probability when they used the terms *likelihood* and the student-generated term *chance* interchangeably with *probability*. As the students gained more experience, they used the term *probability* consistently.

Figures 2.5–2.8 present the work of four seventh-grade students on problem 2.2. Think about the questions in Reflect 2.3 as you inspect these students' work.

Reflect 2.3

Diagrams and explanations provided in response to problem 2.2 by four seventh-grade students—Moses, Kilani, Amelia, and Alana—appear in figures 2.5, 2.6, 2.7, and 2.8, respectively. How does the thinking of these four students differ with respect to the attribute of likelihood or probability?

Do the students' rationales for their solutions indicate additive reasoning or multiplicative reasoning?

What misunderstandings related to ratios and fractions do the students' explanations reveal?

What moves should a teacher make to prompt students to focus on the relevant quantities in their strategies to solve the problem?

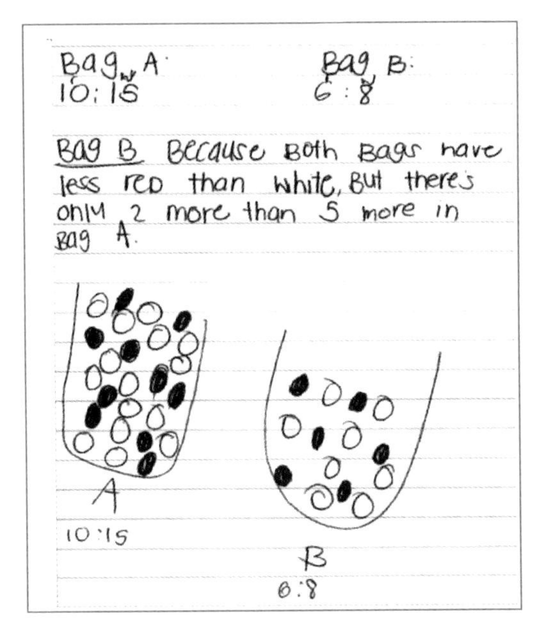

Fig. 2.5. Moses's solution to problem 2.2

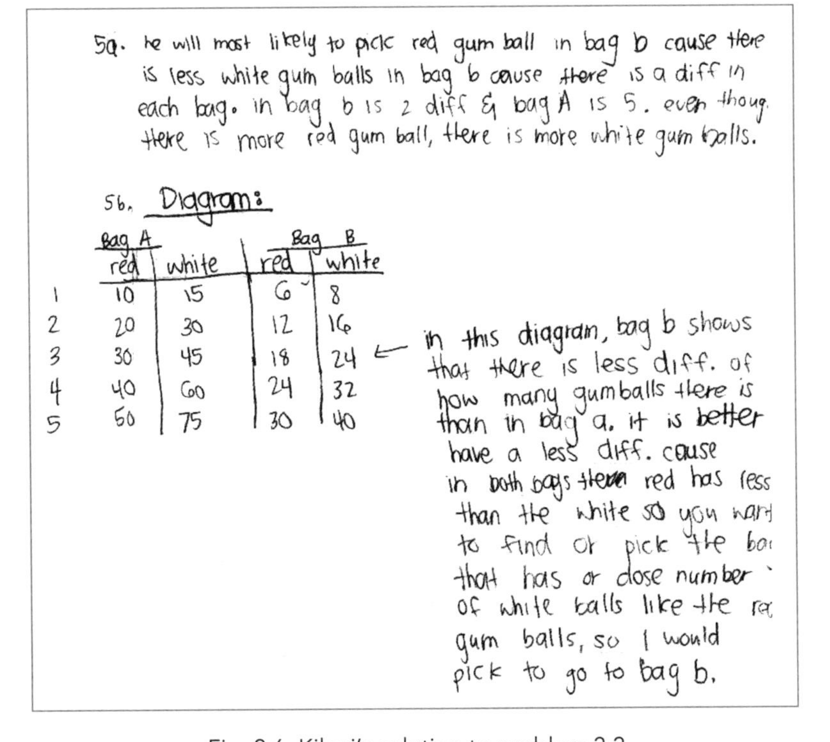

Fig. 2.6. Kilani's solution to problem 2.2

5a.
<u>**Notes**</u>
(gumballs) Bag A- 10 red, 15 white (2/3)
(gumballs) Bag B- 6 red, 8 white (3/4)

Answer: I think that he will most likely pick out a red gumball from Bag B because if you change the amount of gumballs into fractions you would get for Bag A 2/3 and Bag B 3/4. The numerator is the red and the denominator is the white. 3/4 is the larger number so I think that you can get red from that bag. I used a diagram and fractions to help me.

5b.
<u>**Notes**</u>
Make a diagram
Make a diagram

Bag A Bag B

Answer: Each of the bags have the amount of red and white gumballs inside. This helped me because it was a visual aid when I tried to picture the bags. I also could change it into fractions to. So in my head I just simplified the fractions.

Fig. 2.7. Amelia's solution to problem 2.2

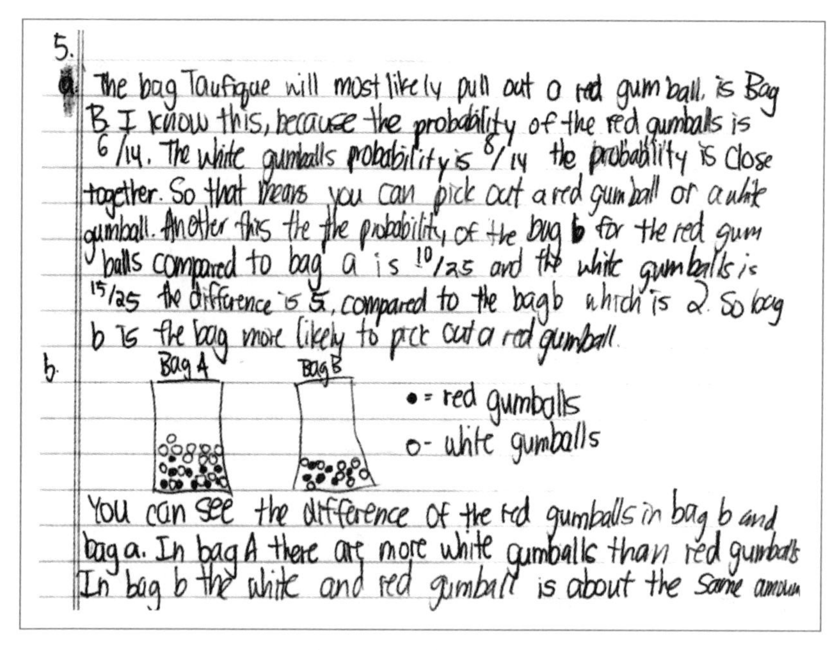

5.
a. The bag Taufique will most likely pull out a red gum ball, is Bag B. I know this, because the probability of the red gumballs is 6/14. The white gumballs probability is 8/14 the probability is close together. So that means you can pick out a red gumball or a white gumball. Another this the the probability of the bag b for the red gum balls compared to bag a is 10/25 and the white gum balls is 15/25 the difference is 5, compared to the bag b which is 2. So bag b is the bag more likely to pick out a red gumball.
b. Bag A Bag B
• = red gumballs
o- white gumballs
You can see the difference of the red gumballs in bag b and bag a. In bag A there are more white gumballs than red gumballs In bag b the white and red gumball is about the same amoun

Fig. 2.8. Alana's solution to problem 2.2

Both Moses and Kilani focused on the *difference* between the numbers of red and white gumballs in each bag. They had the attribute of likelihood in mind, but their rationales do not suggest that they conceptualized probability as the number of possible desired events compared with the number of all possible events. Moses says that he selected bag B "because both bags have less red than white but there's only 2 more" in bag B and "5 more in Bag A."

Kilani likewise based her choice of bag B on the smaller difference between the numbers of red and white gumballs in bag B and explained that she was not misled by the greater number of red gumballs in bag A: "Even though there is more red gum ball [in bag A], there is more white gum balls [in bag A too]."

Moses and Kilani also noted that both bags have more white than red gumballs. Although these students were just two of many who approached the problem by focusing on the differences between the numbers of white and red gumballs in each bag, they were unusual in that they highlighted this condition. Kilani explained, "It is better have a less diff. cause in both bags the red has less than the white so you want to find or pick the bag that has or [a] close number of white gumballs like the red gumballs." She included a diagram showing a series of bags filled with red and white gumballs in the same proportions as bags A and B, and at each stage, the difference between the numbers of red and white gumballs is greater in bag A. Although Kilani's and Moses's conclusions seem to have been built on additive thinking, both students expressed an intuitive or developing notion that when there are fewer gumballs of the color you want to pull, the closer the number of one color is to the number of the other, the more likely your chances are of pulling out a gumball of the desired color.

Amelia also attended to the attribute of likelihood by examining the relationship of red to white gumballs within each bag. But instead of focusing on the differences in the numbers, she used the numbers to create fractions whose numerator and denominator are the number of red gumballs and the number of white gumballs, respectively. For bag A, this process gave Amelia $^{10}/_{15}$, and for bag B, it gave her $^{6}/_{8}$; "so in my head I just simplified the fractions," obtaining $^{2}/_{3}$ and $^{3}/_{4}$. Although the terms that Amelia referred to as "fractions" are in fact ratios written in fraction form, her solution does not make clear what she intended the terms $^{2}/_{3}$ and $^{3}/_{4}$ to represent—the part of the total that is red, the ratio of red to white gumballs, or the probability of pulling out a red gumball. Amelia herself might not have had a clear idea about what the terms represented, but her explanation indicates that she was beginning to think relationally. In other words, she seems to have realized that the attribute of probability is affected by the relationship between two quantities.

Alana, whose work appears in figure 2.8, is the last of the four students whose work is presented on problem 2.2. Alana dealt with the attribute of probability in terms of a fraction representing the number of possibilities for pulling out a red gumball from a bag out of the total number of possibilities for pulling a gumball, red or white, from the bag. She used the relevant quantities to talk about the probability. That is, she compared the desired outcome to all possible outcomes.

Alana made these decisions by looking at the "within" comparison for each bag. She compared the difference between $^6/_{14}$ and $^8/_{14}$ in bag B to the difference between $^{10}/_{25}$ and $^{15}/_{25}$ in bag A, and she chose bag B as the bag from which someone would be more likely to draw a red gumball because its probabilities for drawing red and drawing white gumballs are "close together." In continuing to give her rationale, Alana supported her choice of bag B by observing, "In bag A there are more white gumballs than red gumballs. In bag B the white and red gumball is about the same amount."

Alana's notion of the attribute probability seems to have rested on forming a fraction as a measure of the desired outcome out of all possible outcomes. However, she solved the problem by comparing probabilities within each bag rather than between the bags. Lo and Watanabe (1997) found that the levels of difficulty or challenge presented by the numbers themselves in a problem can affect the methods that students use to solve it, particularly when rational numbers are involved. In solving problem 2.2, Alana found comparing the difference between $^6/_{14}$ and $^8/_{14}$ with the difference between $^{10}/_{25}$ and $^{15}/_{25}$ easier than comparing $^6/_{14}$ with $^{10}/_{25}$.

Solving most probability problems involves the use of proportional reasoning and attention to how the change in each quantity affects the proportion. An example of one such problem follows:

> How does the probability of pulling a white gumball from a bag with 3 white and 2 red gumballs compare with the probability of pulling a white gumball from a bag with 12 white and 8 red gumballs?

Just as some students believe that a road with a 4 percent grade for 0.5 miles is steeper than a road with a 4 percent grade for 0.25 miles, some students believe that the likelihood of pulling a white gumball from the bag containing 12 white and 8 red gumballs is greater than the likelihood of pulling a white gumball from a bag containing fewer gumballs but with white and red gumballs in the same ratio.

Problem 2.3 involves students in working with another relational attribute: speed. In this case, the relevant quantities are numbers of pizzas made and time:

Problem 2.3

In the Fastest Pizza Maker Contest, Keala made three pizzas in five minutes. Casey made four pizzas in nine minutes. If both chefs make pizzas at the same rate as they did in the Fastest Pizza Maker Contest, who do you think will make a pizza faster in a Keala vs. Casey competition?

Nineteen seventh-grade students solved this problem in a unit on ratios and proportions. Figures 2.9–2.11 show the work of three of the students. Use the questions in Reflect 2.4 to guide your examination of the students' work.

Reflect 2.4

Solutions and rationales provided for problem 2.3 by three seventh graders—Chana, Bian, and Makoa—appear in figures 2.9, 2.10, and 2.11, respectively. In what ways did the three students use concepts related to ratio or proportion in solving the problem?

By analyzing the students' work, what can you learn about their conceptions of ratio or proportion?

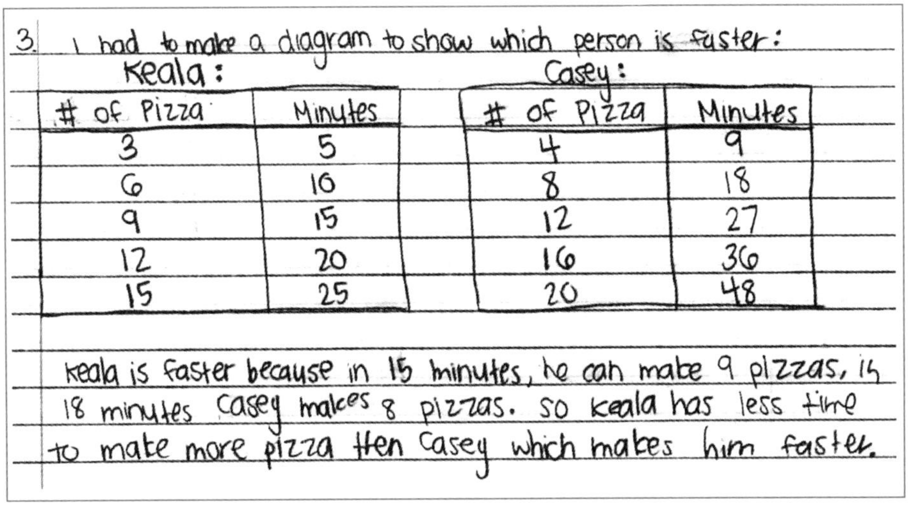

3. I had to make a diagram to show which person is faster:

Keala:

# of Pizza	Minutes
3	5
6	10
9	15
12	20
15	25

Casey:

# of Pizza	Minutes
4	9
8	18
12	27
16	36
20	48

Keala is faster because in 15 minutes, he can make 9 pizzas, in 18 minutes Casey makes 8 pizzas. So Keala has less time to make more pizza then Casey which makes him faster.

Fig. 2.9. Chana's solution to problem 2.3

3.

How I solved the problem?

How I solved the problem for this question was that I made a table and organized it and added up how long it takes for both of the girls to make pizzas into a hour.

Time	Keala	Time	Casey
5	3	9	4
10	6	18	8
15	9	27	12
20	12	36	16
25	15	45	20
30	18	54	24
35	21	63	28
40	24		
45	27		
50	30		
55	33		
60 (1 hour)	36		

Fig. 2.10. Bian's solution to problem 2.3

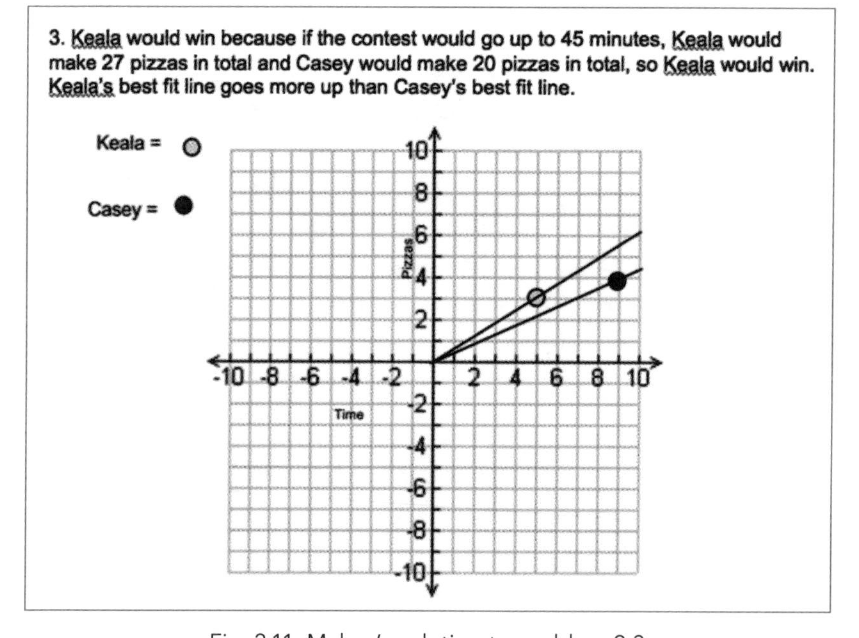
3. Keala would win because if the contest would go up to 45 minutes, Keala would make 27 pizzas in total and Casey would make 20 pizzas in total, so Keala would win. Keala's best fit line goes more up than Casey's best fit line.

Fig. 2.11. Makoa's solution to problem 2.3

All nineteen students chose Keala as the faster pizza maker and were able to identify the quantities relevant to the attribute of speed. However, their explanations indicated that they conceived of the situation in a variety of ways.

Chana, whose work is shown in figure 2.9, iterated composed units (see Lobato and Ellis [2010]) of pizzas and minutes to make tables of equivalent ratios for each pizza chef. From these data, she decided that Keala is faster. Rather than compare the time that each contestant takes to make 12 pizzas—a quantity common to both tables (with Keala taking 20 minutes and Casey taking 27 minutes)—Chana based her comparison of the two chefs on two entries that showed that one chef, Keala, makes a greater number of pizzas in less time than the other chef, Casey. She compared the time that Keala takes to make nine pizzas (15 minutes) with the time that Casey takes to make eight pizzas (18 minutes), and she reasoned, "Keala has less time to make more pizza than Casey which makes him faster."

Bian, whose work appears in figure 2.10, also made tables of equivalent ratios for Keala and Casey, but her goal was to compare the number of pizzas that each chef makes in an hour: "How I solved the problem for this question was that I made a table and organized it and added up how long it takes for both girls to make pizzas into an hour." (The gender of the chefs was fluid for these students.) Bian's method and short explanation about focusing on the number of pizzas per hour suggest that she may have been thinking of a unit rate. Although the problem gives information about the chefs' speeds in minutes, Bian built up increments of pizza-and-time units until her tables were at or close to the hour mark. She settled on this process instead of calculating the number of minutes needed to make one pizza or the number of pizzas made in one minute—a more difficult calculation. Bian did not actually declare who would win the contest, but her table implies that she concluded that Keala is faster and so would be the winner.

Makoa, whose work is presented in figure 2.11, solved the problem with the assistance of a graphing program. The technology enabled him to produce graphs of Keala's and Casey's pizza making on a Cartesian grid from tables of values. The screenshot in the figure shows only a small portion of the Cartesian plane and doesn't show tables of values. However, Makoa was able to use his data to note that at 45 minutes, Keala would have made 27 pizzas and Casey would have made only 20; thus, Keala would win the pizza-making contest. Furthermore, although Makoa did not use the term "rate" or "speed," when he inspected his graphs representing the two rates, he observed, "Keala's best fit line goes more up than Casey's best fit line."

None of the students used sophisticated proportional methods to solve the problem. However, all of them incorporated aspects of proportional reasoning in their solutions. All three students used a "building up" strategy to create data for composed units—number of pizzas to time. Chana and Keala created tables, and Makoa used software to create graphs that he realized would continue beyond what would fit on his screen. In using the data to determine who would win the pizza-making contest, each of these students chose a place in the data where it was easy for them to make a "more or less" comparison. They might not have realized it, but in using varying data to make the comparison, they assumed that the measure of the attribute (speed of pizza making) remained constant as the quantities (the composed units) changed within the data set for each chef.

Summarizing Pedagogical Content Knowledge to Support Essential Understanding 3

Teaching the mathematical ideas in this chapter requires a specialized knowledge related to the four components presented in the Introduction: learners, curriculum, instructional strategies, and assessment. The four sections that follow summarize some examples of these specialized knowledge bases in relation to Essential Understanding 3, which relates to forming a ratio as the measure of a real-world attribute. Although we separate the four components of pedagogical content knowledge to highlight their importance, we also recognize that they are connected and support one another.

Knowledge of learners

Teachers are often surprised when students do not apply concepts and skills to real-world problems although they have received instruction in these same concepts and skills. When Lobato and Thanheiser (2002) gave students problems involving slope (presented in the context of the steepness of a wheelchair ramp) and speed (set in the context of how fast a person walks), they found that the students were unable to isolate the relevant attribute, even though they had studied these topics. This inability prevented them from formulating an appropriate ratio to measure the attribute. The students appeared to be unable to move from decontextualized problems involving slope and speed to contextualized problems, as called for in the Common Core State Standards for Mathematics (National Governors Association Center for Best Practices and Council of Chief State School Officers [NGA Center and CCSSO] 2010, p. 6):

> [Students] bring two complementary abilities to bear on problems involving quantitative relationships: the ability to *decontextualize*—to abstract a given situation and represent it symbolically and manipulate the representing symbols as if they have a life of their own, without necessarily attending to their referents—and the ability to *contextualize*, to pause as needed during the manipulation process in order to probe into the referents for the symbols involved.

This chapter has shown that students' difficulties in identifying relevant attributes in real-world problems can stem from a variety of causes, including perspectives or assumptions that the students bring to the problem from prior experience with the situation or their lack of experience in reasoning quantitatively. Giving students tasks such as problem 2.1, which requires that they focus on relevant attributes, can build awareness of the fact that different contexts may require different types of quantitative reasoning (Lobato and Siebert 2002).

Chapter 2 has also shown that students may obtain the correct answer to a problem but approach it in different ways. Or they may obtain a correct result that is based on partially (or totally) flawed reasoning. In solving problem 2.2, most students arrived at the right conclusion—that Taufique is more likely to draw a red gumball from bag B than from bag A—by using the differences of "within" ratios of red gumballs to white gumballs in the two bags rather than a comparison of the probabilities of pulling a gumball of the desired color from the two bags. The fact that the students were thinking about the problem and solving it in this way might not have been evident if the problem had not asked students to draw a diagram that represented their solution method and explain it.

Knowledge of curriculum

This book consistently calls attention to the importance of engaging students in tasks that support the development not only of procedural skill but also of deep conceptual understanding. Such tasks have certain characteristics. Not every good task will have them all, but every task in which a student engages should promote growth in skill and understanding. The tasks in this chapter are designed to serve several different purposes. Problem 2.1 prompts students to consider and reflect on their conceptions of measurable attributes of a polygon. The open-ended nature of the problem invites discussion in which students can express their ideas, hear others' points of view, and clarify their thinking.

Problem 2.2 is designed to bring to the surface common misunderstandings that students may have about ratios and probability. Working with the problem can help students see that greater numbers of items—the 10 red gumballs and 15 white gumballs in bag A as compared with the 6 red gumballs and 8 white gumballs in bag B—do not necessarily result in a greater ratio or probability. A problem that switches the color of interest from red to white might be a good follow-up task. Could students have used the "within" differences of red and white gumballs in the same way if they were asked from which bag, bag A or bag B, someone is more likely to pull out a white gumball?

All three problems yielded student work that show that the tasks are challenging but accessible to students. In other words, students had some way to begin solving the problem, even if all their reasoning was not yet fully formed.

Having a good understanding of the curriculum will enable you to select or adapt tasks that address your students' needs. Being able to analyze a task's intent can help you to tailor the curriculum to the teaching situation while at the same time maintaining alignment with content expectations expressed in local and national standards.

Knowledge of instructional strategies

Another message that pervades this book is that students interpret problems differently and approach their solutions differently—no surprise there, of course! Recall Evangeline's notes on the class discussion about problem 2.1 (fig. 2.4), reflecting her reconsideration of the problem and possible ways to solve it.

Implementing instructional strategies that take advantage of these differences promotes students' development of deeper and more robust conceptual understanding. Allowing students to work collaboratively on some problems, having them present their solutions to the class, and discussing solutions that they obtained by means of various approaches are all instructional strategies that you can use to elicit your students' different points of view. These are some of the instructional strategies that can help your students make mathematical connections among multiple models of problem situations, pose questions to others, and assess their own understanding of the mathematics.

Knowledge of assessment

If you are to help your students think more deeply about the mathematics that they are encountering, the assessments that you use to measure their thinking need to go beyond determining whether they have gotten the right answer. Requiring your students to include explanations and diagrams in their responses can reveal levels of understanding and areas of misunderstanding that may otherwise be hidden by a correct answer. You have probably discovered for yourself that writing an explanation of how you solved a problem can often prompt reflection that strengthens your own understanding. Yet, for most students, writing in mathematics class is a foreign activity. Students have not had many opportunities to write about math, and they often find it difficult to articulate what they are thinking in solving a problem. This is true for students regardless of their level of success in mathematics. Although having all students write with the fluidity that Thomas displayed (see fig. 2.2) may not be an attainable goal, you can begin to support your students in building their capacity to write about mathematics.

Conclusion

The samples of students' work provided in this chapter point to mathematics that teachers must understand if they are to understand their students as individual learners, design learning opportunities for them, make critical instructional decisions, and interpret students' responses to assessments in meaningful ways. As a teacher, you need to understand ways in which your students attend to, or may attend to, measurement of attributes expressed as relationships when they are reasoning to solve problems involving ratios and proportions.

The examples in this chapter have demonstrated that it is possible for students to get the correct answer to a problem without truly understanding the attribute of critical importance. Not all real-world situations calling for the application of mathematics, such as probability problems, speed problems, or steepness problems, present the same degree of difficulty. Why were the students closer to focusing on the correct attribute in the case of the pizza-making contest (problem 2.3) than they were in the draw-a-red gumball situation (problem 2.2)? Had they simply had more experience with speed than with chance? Such matters have important implications for introducing related topics that call for students to apply an understanding of ratio and probability.

Chapter 3
Focusing on Equivalent Ratios

Essential Understanding 7

Proportional reasoning is complex and involves understanding that—

- equivalent ratios can be created by iterating and/or partitioning a composed unit;
- if one quantity in a ratio is multiplied or divided by a particular factor, then the other quantity must be multiplied or divided by the same factor to maintain the proportional relationship; and
- the two types of ratios—composed units and multiplicative comparisons—are related.

The understanding that students develop in grades 3–5 about fractions and multiplicative relationships can give them a solid foundation on which to build an understanding of ratios and proportions in grades 6–8. Thompson and Saldanha (2003) give numerous examples of relationships that are important for students to grasp between fractions and multiplicative reasoning, on the one hand, and ratio and proportion, on the other. In particular, they state, "Students gain considerable mathematical power by coming to understand fractions through a scheme of operations that express themselves in reciprocal relationships of relative size" (pp. 107–8). They explain that knowing about these relationships is useful in solving a related type of rate problem that is often difficult for algebra students. Expressing basically the same idea in a slightly different way, Chval, Lannin, and Jones (2013) state, "Work with fractions in grades 3–5 provides an entry point for reasoning about various mathematical topics, such as ratio, rate, proportion, and percentage" (p. 122). By building on the understanding of equivalent fractions that your students have and the multiplicative reasoning on which their understanding rests, you can move them toward a rich understanding of ratios, rates, and proportions.

Working toward Essential Understanding 7

To examine students' reasoning related to equivalent fractions, and to consider how this thinking may relate to reasoning about equivalent ratios, we gave problem 3.1 to seventh-grade students and also asked them to justify their response.

Problem 3.1

Are the fractions $^{6}/_{9}$ and $^{10}/_{15}$ equivalent?

As observed by Lobato and Ellis (2010), "Mathematics uses several conventional notations to represent ratios" (p. 19). For example, in comparing the ratio of two quantities, you could report equivalent ratios as "2:5," "2 to 5," or "$^{2}/_{5}$." Although there is usually no confusion about the fact that "2:5" and "2 to 5" represent ratios, at different times "$^{2}/_{5}$" might be thought of as a ratio or as a fraction. Those using this form usually attempt to indicate which they intend.

Five students' responses to problem 3.1 appear in figures 3.1–3.5. We chose these responses not because they represent all possible aspects of student thinking but because they demonstrate a range of connections with ideas involved in proportional thinking. These five responses also illustrate how students' responses can be used to probe their thinking about ratios and expand their ability to reason about them when the students engage in problems that involve ratio settings rather than fraction settings. Consider the questions in Reflect 3.1 as you inspect the students' responses shown in the figures.

Reflect 3.1

Figures 3.1–3.5 show five seventh-grade students' responses to problem 3.1.

In what ways do the students' responses indicate the complex nature of reasoning involving equivalent fractions?

In what ways can this reasoning be useful in settings involving ratios instead of fractions?

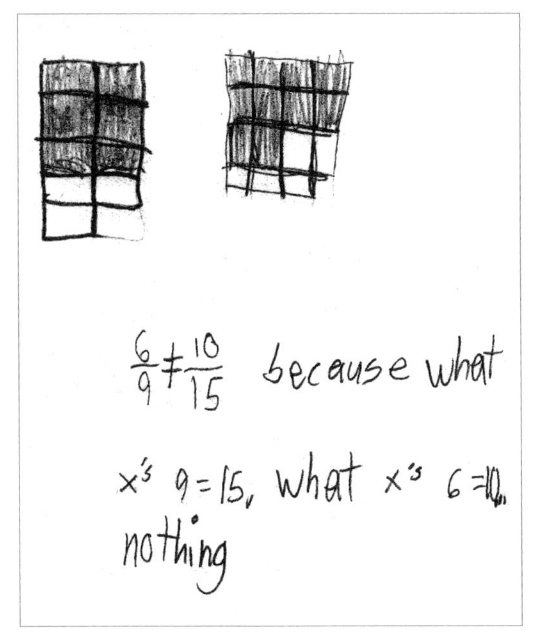

Fig. 3.1. Sean's response to problem 3.1

No because $\frac{10}{15}$ can't be Simplafyed to $\frac{6}{9}$

Fig. 3.2. Ernie's response to problem 3.1

Explain your reasoning and support your answer.

yes, Becase you can multiply the first fraction by 5, and the 2nd Fraction By 3 and they well equal.

Fig. 3.3. JQ's response to problem 3.1

$\frac{6}{9} \times \frac{10}{15} = \frac{90}{90}$ Yes because they = a number over its self

Fig. 3.4. Chrystal's response to problem 3.1

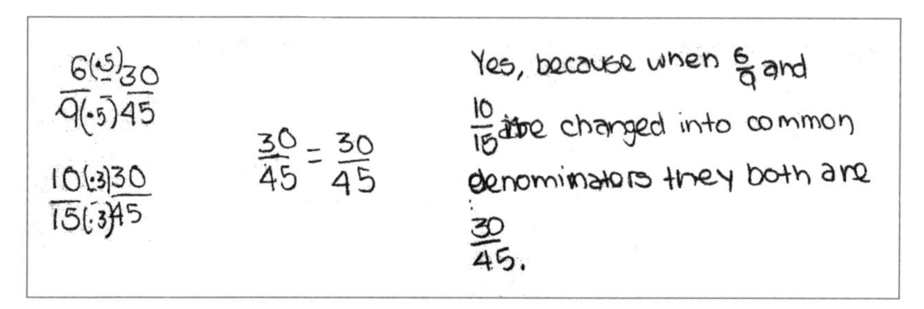

Fig. 3.5. Karin's response to problem 3.1

Sean's response, shown in figure 3.1, seems to indicate that Sean used a multiplication strategy to determine whether the fractions $^6/_9$ and $^{10}/_{15}$ are equivalent. She decided that they are not equivalent because she could not use the same whole number to multiply 6 to obtain 10 and to multiply 9 to obtain 15. Sean was trying to use the idea of multiplying "across" numerators and denominators. The model that she attempted to apply—diagrams composed of shaded and unshaded squares—did not prove useful, but had she arranged the squares as in the diagrams in figure 3.6, perhaps she would have seen the equivalence of the fractions $^6/_9$ and $^{10}/_{15}$. Such diagrams are often used in a "building up" approach to demonstrate the equivalence of two fractions or ratios.

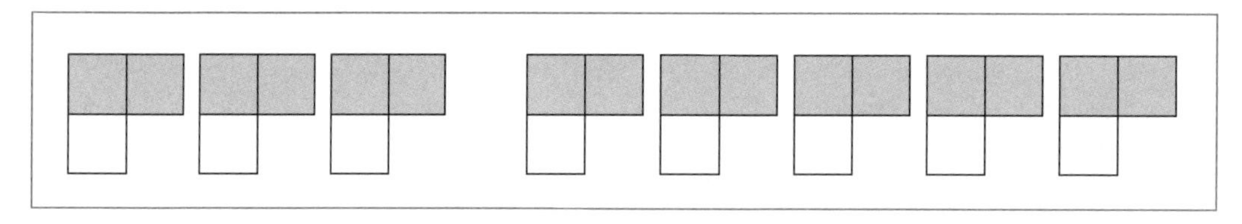

Fig. 3.6. A "building up" approach to equivalent fractions or ratios

Similarly, plotting the points $^6/_9$ and $^{10}/_{15}$ on parallel number lines, as shown in figure 3.7, is useful in determining whether the two fractions name the same point and thus are equivalent. According to the Common Core State Standards for Mathematics (CCSSM; National Governors Association Center for Best Practices and Council of Chief State School Officers [NGA Center and CCSSO] 2010), students in grade 3 should be able to "represent a fraction $^{a}/_{b}$ on a number line diagram by marking off *a* lengths $^{1}/_{b}$ from 0" (p. 24). Figure 3.7 illustrates the use of parallel number lines to make the equivalence of fractions visible. The mark for the sixth subunit on the top line indicates $^{6}/_{9}$, and it lines up with the mark for the tenth subunit on the bottom line, which indicates $^{10}/_{15}$, thus demonstrating the equivalence of $^{6}/_{9}$ and $^{10}/_{15}$. Also implicit is the fact that each of these fractions represents $^{2}/_{3}$; although not labeled, the third subunit mark on the top line (indicating $^{3}/_{9}$) lines up with the fifth subunit mark on the bottom number line (indicating $^{5}/_{15}$).

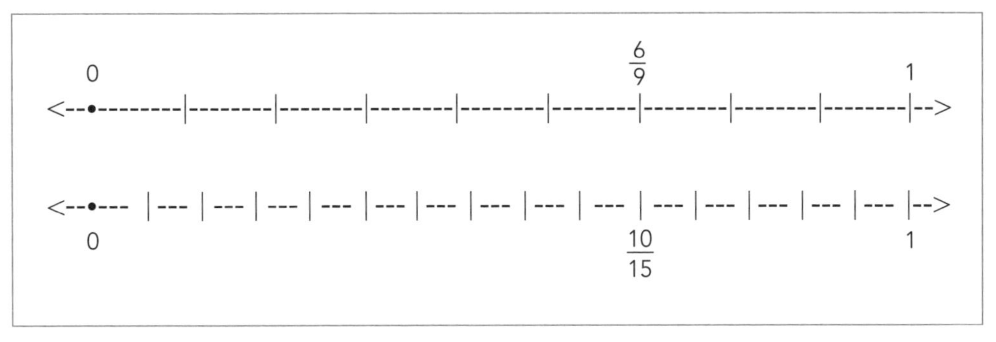

Fig. 3.7. Demonstrating the equivalence of the fractions $^{6}/_{9}$ and $^{10}/_{15}$ on parallel number lines

Ernie's response to problem 3.1 (see fig. 3.2) is similar to Sean's, except that Ernie assumed that if the fractions $^{6}/_{9}$ and $^{10}/_{15}$ are equivalent, he would be able to find a whole number that he could use to divide 10 and 15 by to obtain 6 and 9, respectively. Consider Sean's and Ernie's responses more closely by responding to the question in Reflect 3.2.

Reflect 3.2

How do you suppose that Sean and Ernie, whose work is shown in figures 3.1 and 3.2, respectively, would have responded to the following problem—a variation on problem 2.2?

> **You have two bags containing blue and red gumballs. Bag A contains 6 blue gumballs and 9 red ones, and bag B contains 10 blue gumballs and 15 red ones. From which bag would you have the better chance of selecting a blue ball?**

Thinking similar to that modeled in figure 3.6 could be used to show that for every 2 blue balls *in either bag, A or B,* in the problem in Reflect 3.2, there are 3 red balls. Hence, you would be equally likely to pick a blue ball from either bag.

Arguably, JQ's work, which appears in figure 3.3, illustrates that he understood the usefulness of multiplying by a fractional form of 1, but he did not articulate the idea correctly or give an example to clarify what he was suggesting. If JQ had been given the problem in Reflect 3.2, he might have demonstrated the meaning and value of his approach. By thinking in terms of ratios and using the model shown in figure 3.6, he could have "built up" the ratio of 6 blue gumballs to 9 red ones ($^{6}/_{9}$) in bag A five times to produce the ratio of 30 blue gumballs to 45 red ones ($^{30}/_{45}$), while preserving the invariant ratio of 6 to 9. If he had done so, he would have demonstrated his understanding that "the ratio of one quantity to the other is invariant as the numerical values of both quantities change by the same factor" (Lobato and Ellis 2010, p. 11). Similarly, in working with the ratio of 10 blue balls to 15 red balls ($^{10}/_{15}$) in bag B, he could have built up this ratio 3 times, also producing 30 blue balls to 45 red balls ($^{30}/_{45}$) while preserving the invariant ratio of 10 to 15. By this method, he might have found that he could build up each relationship to a ratio of 30 blue balls to 45 red balls, thus verifying the equivalence of $^{6}/_{9}$ and $^{10}/_{15}$ that he stated in his response to problem 3.1.

In a casual examination, the sign between $^{6}/_{9}$ and $^{10}/_{15}$ in Chrystal's solution, shown in figure 3.4, might appear to be a multiplication symbol. However, it is almost certainly an indication of Chrystal's application of a form of the cross multiplication procedure. Her work suggests that that she obtained $^{90}/_{90}$ by multiplying 9×10 and 6×15. Although a similar procedure is often used in work with proportional relationships, Chrystal provided no context or rationale to help others make sense of her computational actions.

Karin's response, shown in figure 3.5, combines ideas shared by other students and arguably provides an appropriate explanation. Karin multiplied both the numerator and the denominator of $^{6}/_{9}$ by 5 to obtain the equivalent fraction $^{30}/_{45}$, and then she multiplied both the numerator and the denominator of $^{10}/_{15}$ by 3 to obtain the equivalent fraction $^{30}/_{45}$, thereby demonstrating the equivalence of the original fractions, $^{6}/_{9}$ and $^{10}/_{15}$. Perhaps this is the process that JQ was also attempting to articulate. Students using similar logic would probably also be able to use a "building up" strategy to demonstrate that the chance of drawing a blue ball from bag A is equal to that of drawing a blue ball from bag B in the situation presented in Reflect 3.2.

These five samples of student work illustrate the fact that when seventh-grade students are given a problem that asks them to determine whether two fractions are equivalent, they demonstrate thinking that relates to Essential Understanding 7. That is, they show some knowledge that is useful for understanding and creating equivalent ratios by iterating or partitioning a composed unit and maintaining equivalent ratios by multiplying or dividing both quantities of a ratio by the same factor. However, in inspecting their work and attempting to discern their thinking, teachers need to take care to differentiate between the way in which students see equivalence when dealing with fractions and the way in which they see it when dealing with ratios. For example, Karin's process allowed her to see $^{6}/_{9}$ and $^{10}/_{15}$ as equivalent fractions, but she was not thinking about multiplying quantities in a ratio by the same factor. Yet, if she had encountered these values in the context given in Reflect 3.2, she might have thought about what she was doing as multiplying quantities in ratios by the same factor. In essence, thinking about the numbers in context is important.

Other complex issues make proportional reasoning difficult for middle-grades students. Particular types of questions can both help reveal students' thinking and engage students in thinking that will strengthen their proportional reasoning skills. Reflect 3.3 provides a useful entry point to a discussion of these types of questions.

Reflect 3.3

Consider the two representations, J and K, in figure 3.8. Are these two representations equivalent?

What information might you find helpful before answering the question?

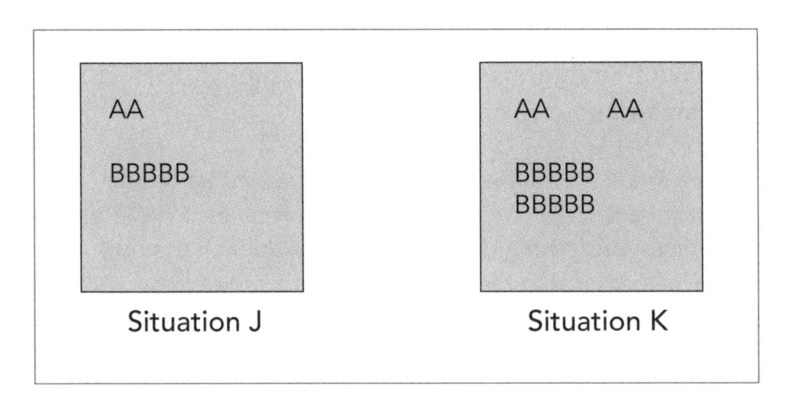

Fig. 3.8. Representations J and K

Unless those considering representations J and K have reached an understanding of what these situations represent and what a comparison of them involves, they will find it difficult to decide whether the situations are equivalent. If the relationship between A's and B's is the focus of the comparison of the situations, then context is significant because different contexts may affect interpretations of the relationship. Two scenarios are presented and discussed below to highlight these points. Both offer possible contexts for the representations in figure 3.8.

> **Scenario 1:** J and K are siblings who earn money in the summer by mowing lawns in their neighborhood. Each has particular lawns that he or she mows, and all the lawns have the same area and take the same amount of time to mow. To help them organize their work, the siblings make weekly charts for themselves, using A's to represent lawns that they have mowed and B's to represent lawns yet to be mowed.

Having a context for figure 3.8 that relates specifically to lawn mowing helps focus attention on important characteristics of a situation. Suppose that the figure represents J's and K's charts of lawns mowed (A's) and lawns yet to be mowed (B's) in a particular week. Does this information permit determining whether the situations are equivalent?

Although the total number of lawns, number of lawns mowed, and number of lawns yet to mow are different for J and K, particular ratios within the setting are equivalent. By using either a building-up or a partitioning strategy, students could determine that the ratios of lawns moved to lawns not yet mowed for J and K (that is, 2 to 5 and 4 to 10, respectively) are equivalent. They could also determine other equivalent ratios, such as the ratios of lawns mowed to total lawns scheduled for J and K (specifically, 2 to 7 and 4 to 14), or lawns yet to be mowed to total lawns scheduled for J and K (5 to 7 and 10 to 14).

This look at scenario 1 demonstrates the complexities of examining a situation and making claims regarding equivalent ratios with respect to its context. Scenario 2 locates the information in figure 3.8 in a different context, further illuminating the complexities of proportional reasoning.

> **Scenario 2:** J and K are siblings whose birthdays are just one day apart. Both are planning small parties where they will serve cupcakes. J and K are making charts to help them get everything that they need. On their charts they are using A's to represent people invited to the party and B's to represent cupcakes that they will offer at their respective parties. That is, J has invited 2 people and will provide 5 cupcakes, and K has invited 4 people and will provide 10 cupcakes.

Note that the information in the party setting is different in subtle ways from the information in the lawn setting. From a proportional perspective, statements like the following express how the situations for J's and K's parties can be considered equivalent:

- Each sibling plans to order 5 cupcakes for every 2 people, allowing $2^1/_2$ cupcakes per person at the party for each sibling.
- Each sibling's party will have $^2/_5$ person per cupcake.
- $^2/_5 = {}^4/_{10}$, using the building-up method.
- The party for sibling K will have twice as many cupcakes but also twice as many people as the party for sibling J.

Depending on definitions in your mathematics curriculum or your students' prior experiences, your use of "per" or "for every" may signify to them that this relationship is a *rate* or *measurement*. Phrasing that you might use to describe some relationships in scenario 1 has a companion that you might apply to a relationship in scenario 2, but phrasing that you might use to describe other relationships in scenario 1 has no counterpart that you could apply to a relationship in scenario 2. For example, in talking about the lawn-mowing setting in scenario 1, you might say that each sibling has 5 lawns yet to mow *for every* 2 lawns mowed. The companion statement in the case of scenario 2 is that each sibling's party will have 5 cupcakes *for every* 2 people. However, saying in relation to the lawn setting in scenario 1 that J and K have mowed 2 lawns for every 7 lawns that they have scheduled has no companion in the party setting in scenario 2.

Figure 3.9 returns to the use of parallel number lines to represent the equivalence of fractions or ratios—the model illustrated in 3.7. The representation in figure 3.9 shows a different approach and way of thinking about the relationship of equivalence between $^3/_4$ and $^6/_8$. Examine the figure, and then respond to the question in Reflect 3.4.

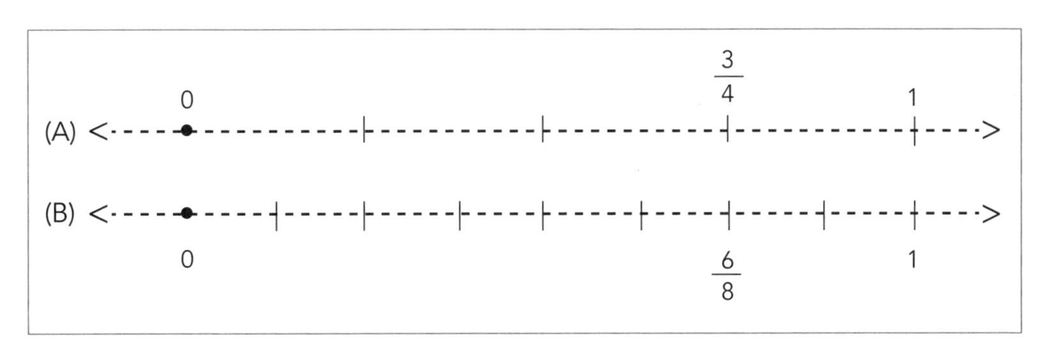

Fig. 3.9. Showing equivalence by using parallel number lines

Reflect 3.4

Consider the use of parallel number lines to represent equivalence shown in figure 3.9. Which would be more appropriate to use, partitioning or multiplication, to explain why 3/4 = 6/8 with respect to this representation?

The marks and segments in figure 3.9 show that understanding equivalent fractions, and hence equivalent ratios, and working with them, involve further complexity. Can any claim of equivalence be made for the fraction $^3/_4$ located in representation A and the fraction $^6/_8$ located in representation B? Representation A shows 4 subunits, and representation B shows 8 subunits. The third mark of the subunits in representation A lines up with the sixth mark of the subunits in representation B. In general, given these fractions, you might claim that $^3/_4$ is equivalent to $^6/_8$ because you can multiply $^3/_4$ by $^2/_2$ and obtain $^6/_8$. Likewise, in the lawn and cupcake examples (scenarios 1 and 2), you might justify the claim that $^2/_5$ is equivalent to $^4/_{10}$ by building up $^2/_5$ to $^4/_{10}$ by doubling the 2 and the 5 to obtain the 4 and the 10.

However, the demonstration of the equivalence of $^3/_4$ and $^6/_8$ in the number line example in figure 3.9 is not as simple as multiplying both the numerator and denominator of $^3/_4$ by 2. Although such arithmetic is accurate, representation B does not show a doubling (or multiplying) of the length of the subunits (3 out of 4 partitions of 1) in representation A. The number line example is, in effect, the opposite of the lawn example. That is, in figure 3.9, each of the 3 subunits in representation A is *partitioned* into two smaller subunits, and doing so creates *twice as many* (smaller) subunits in representation B (with $^6/_8$ marking off 6 of 8) as in representation A (with $^3/_4$ marking off 3 of 4). This illustrates the importance of focusing on the relationship among units used in the context under consideration.

By contrast, showing $^6/_8 = {}^3/_4$ would reverse the process. Using the same parallel number lines, you would start with the representation for $^6/_8$ and work from it to justify that the representation for $^3/_4$ shows the same point. In general, given the task of justifying that $^6/_8$ is equivalent to $^3/_4$, you might say, "The 6 and the 8 can both be divided by 2 to obtain $^3/_4$." However, in the number line representation, you would see two of the subunits in representation B being consolidated into a larger subunit on representation A. That is, you would focus on the fact that the subunits in representation A are *twice as large* as the subunits in representation B, so the point marking 6 subunits of 8, or $^6/_8$, is associated with the same point as that associated with 3 subunits of 4, or $^3/_4$. In other words, because 6 of the subunits of 1 in representation B are equivalent to 3 of the subunits of 1 in representation A, and 8 subunits of 1 in representation B are equivalent to 4 subunits of 1 in representation A, you know that $^6/_8$ and $^3/_4$ represent the same point on a number line and are the same value. Because a fraction is a representation of the value of a ratio, students' understanding of this setting has important implications for their understanding of ratios. Consideration of this setting also demonstrates the usefulness of the building-up and partitioning strategies for examining either equivalent ratios or fractions.

These ideas related to fractions also need to be compared and contrasted with the structures and meanings of ratios in proportional settings. The use of a number line to determine equivalence reflects a different approach from the one most likely to be used in deciding whether 3 oranges for 4 people is equivalent to 6 oranges for 8 people. Most people would use a composed-units approach in deciding this question or in examining, for instance, the equivalence of 3 parts of orange juice to 4 parts of water and 6 parts of orange juice to 8 parts of water. In such settings, adding ratios is different arithmetically from adding fractions.

For example, consider a situation in which for a group of four people the ratio of oranges to people is 3 to 4. Suppose that in a separate group, the ratio of oranges to people is the same, 3 to 4. If these two groups of people merged with their respective ratios, then what would be the ratio of oranges to people in the resulting group? If we assume that each group has exactly 3 oranges and 4 people, the combined group will have 6 oranges and 8 people. Hence, the ratio of oranges to people will be 6 to 8, or equivalently 3 to 4. The merging of the groups did not change the ratio; that is, the ratio was invariant. However, the fractions associated with the ratios cannot be added in the conventional way. In fact, in this situation, if one of your students wrote

$$\frac{3}{4}+\frac{3}{4}=\frac{6}{8},$$

you would suspect either that the student did not understand fractions or that she or he had a keen understanding of ratios but was manifesting it in unfortunate and incorrect mathematical notation.

In a problem context, if a setting requires 3 oranges for 4 people and there are 8 people, problem solvers can determine the number of oranges needed by looking at $8 \div 4 = 2$. That is, because there are twice as many people, twice as many oranges (6) are needed. Any setting requires problem solvers not only to think about multiplying or dividing to create equivalent fractions but also to keep in mind the relationship between the quantities involved. Although the fraction associated with a ratio is a number, the values of the numerator and denominator are quantities, or units, that are related in some manner, and maintaining that invariant relationship is crucial in a proportional setting.

Problem 3.2, which is set in a context that involves both fractions and ratios, was given to students in grades 5–8 to solve. Reflect 3.5 poses a question about students' possible approaches to this problem. Read the problem with this question in mind.

Reflect 3.5

How can students use sense making rather than rules or equations in formulating solutions to Problem 3.2?

Problem 3.2

Jonnine had a board. She cut and used $^2/_5$ of the board for bracing. She measured the piece used for bracing and found it to be $^3/_4$ foot long. How long was the original board?

The student work on problem 3.2 in figures 3.10 and 3.11 offers insight into how students use aspects of Essential Understanding 7 to make sense of situations involving ratios and fractions. Although students can solve this problem in many ways, the two students whose work is shown combined multiplicative, additive, and building-up processes in using proportional reasoning. For additional discussion of students' work on, as well as teachers' conceptions of, the task presented here as problem 3.2, see Sjostrom, Olson, and Olson (2010).

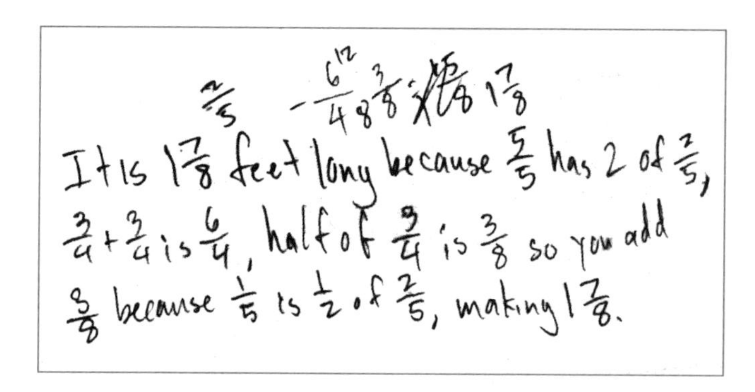

Fig. 3.10. Keo's response to problem 3.2

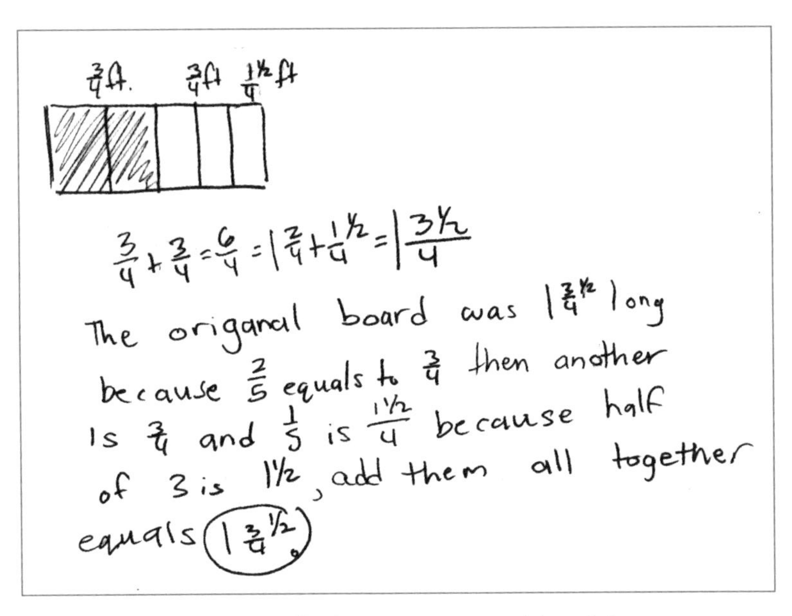

Fig. 3.11. Elise's response to problem 3.2

Keo's response shows a combination of additive and building-up methods, along with a use of other properties of ratios. Most notably, he recognized the value of using $^{5}/_{5}$ for the length of the original board. He understood that there are two $^{2}/_{5}$'s in $^{5}/_{5}$. Knowing that $^{2}/_{5}$ board was $^{3}/_{4}$ foot long, he built two $^{2}/_{5}$'s, which gave him a corresponding length of $^{6}/_{4}$ foot of board. He then used the proportional relationship between $^{2}/_{5}$ and $^{3}/_{4}$ to reason "$^{1}/_{5}$ is $^{1}/_{2}$ of $^{2}/_{5}$" ($^{2}/_{5} : {}^{3}/_{4} = {}^{1}/_{5} : {}^{3}/_{8}$) to find the length associated with the final $^{1}/_{5}$ of the board and combined $^{6}/_{4}$ and $^{3}/_{8}$ to determine the length of the board.

Elise's response is similar to Keo's, both in the additive and building-up methods used to determine that two $^{2}/_{5}$'s of the board would be $^{6}/_{4}$ feet long and in the recognition of the need to determine the foot value associated with $^{1}/_{5}$ of the board. Differences arise in the manner in which Elise determined and used the fractional values. Elise used proportional reasoning to determine that if $^{2}/_{5}$ of a board was $^{3}/_{4}$ feet long, then $^{1}/_{5}$ of the board must be

$$\frac{1\frac{1}{2}}{4} \text{ feet long } \left(\frac{2}{5} : \frac{3}{4} = \frac{1}{5} : \frac{1\frac{1}{2}}{4}\right).$$

The fact that Elise was able to use this thinking and arrive at a complex mixed-number fraction is significant because at this stage her reasoning was seemingly numerically independent, contextually powerful, and generalizable. That is, if the numerical presentation of the problem had been different, with the problem giving, say, $^{2}/_{29}$ of the board as $^{3}/_{43}$ feet long, Elise's reasoning might well have maintained independence from the numerals in the denominator: she might readily have understood

$$\frac{1}{29} \text{ of the boars as } \frac{1\frac{1}{2}}{43} \text{ feet long.}$$

In other words, if students are thinking in terms of fractions, the context of the situation might prompt them to divide the numerators because of the relationships involved. Alternatively, if they are thinking in terms of ratios, they might reason that

$$\frac{2}{29} : \frac{3}{43} = \frac{1}{29} : \frac{1\frac{1}{2}}{43}$$

by using the property that if one quantity in a ratio is multiplied or divided by a particular factor, then the other quantity must also be multiplied or divided by the same factor to maintain the proportional relationship.

Summarizing Pedagogical Content Knowledge to Support Essential Understanding 7

Teaching the mathematical ideas in this chapter requires specialized knowledge related to the four components presented in the Introduction: learners, curriculum, instructional strategies, and assessment. The four sections that follow summarize some examples of these specialized knowledge bases in relation to Essential Understanding 7. Although we separate them to highlight their importance, we also recognize that they are connected and support one another.

Knowledge of learners

As demonstrated by the students' work on problem 3.1, which asks whether $^{6}/_{9}$ and $^{10}/_{15}$ are equivalent, students have both misconceptions about evaluating the equivalence of fractions and innovative ways to determine their equivalence. On one hand, Sean and Ernie (see figs. 3.1 and 3.2, respectively) did not think $^{6}/_{9}$ and $^{10}/_{15}$ are equivalent, because they could not readily see a multiplicative relationship—that is, a whole number multiplier—that was the same between 6 and 9 and 10 and 15. On the other hand, JQ and Karin (see figs. 3.3 and 3.5, respectively) saw that each of the fractions could be changed to equivalent fractions with the same numerators and denominators—a process that mirrors a building-up strategy for finding equivalent ratios. Later in the chapter, Keo's and Elise's responses (see figs. 3.10 and 3.11, respectively) to problem 3.2 about Jonnine's board showed the value of having students use the context of a problem to demonstrate their emerging ability to reason about ratios and proportions.

Knowledge of curriculum

The emphasis in Chapter 3 has been on how middle-grades students "expand the scope of problems for which they can use multiplication and division to solve problems, and they connect ratios and fractions," as CCSSM asserts in introducing the standards for grade 6 (NGA Center and CCSSO 2010, p. 39). Often, without intending to do so, textbooks suggest by the way in which they organize their content that many ideas related to ratio and proportional reasoning can be segmented into discrete instructional and learning parts. Although CCSSM situates extensive work on ratios and proportions in grades 6 and 7, the concepts presented in this chapter should not be considered as standing alone but as foundational for many other mathematical topics, such as relationships in tables, proportionality in scaling, similarity in dilations, slopes of lines (both through the origin and not through the origin), regression equations, and proportional equations.

Reflect 3.2 presents a contextualized variation on problem 3.1. Bag A has 6 blue gumballs and 9 red ones. Bag B has 10 blue gumballs and 15 red ones. From which bag is the chance of drawing a blue gumball greater? The discussion related to Reflect 3.2 highlights the complexity of the reasoning involved in comparing fractions and ratios and the importance of contextual situations in developing understanding of, and reasoning about, ratios and proportions. It also shows how various ratio contexts can affect your own and your students' conceptualizing of rates and use of language for comparing two quantities. The thinking that working in a context requires goes beyond the thinking often associated with ratios and rates contexts where simple multiplication and division can be used to answer questions—for example, in working with tables of values representing money associated with hours of work ($12 per hour, for instance).

Knowledge of instructional strategies

You can employ many strategies to help your students. When they manipulate numbers to express relationships in ratios, you should ask them to say what they envision in the ratios that they have identified. For example, in the cupcake setting, you might ask, "What are you seeing when you give the ratio of 2 to 5?" Similarly, if you ask whether $^2/_5$ is equivalent to $^4/_{10}$, your students should be able to create a context in which this equivalence makes sense.

Knowing strategies for demonstrating equivalent fractions is not only fundamental to showing the equivalence of ratios and for building proportional reasoning skills but also useful for reasoning in context. Just as using benchmark fractions to compare $^3/_7$ and $^5/_8$ ($^3/_7$ is less than $^1/_2$ whereas $^5/_8$ is larger than $^1/_2$), the building-up strategy is also useful for understanding comparisons involving ratios that are not equivalent. For example, suppose that your students are working with a problem about someone who is reconstituting orange juice. If the problem says that she starts with 3 cans of orange juice and 4 cans of water (3:4) and adds to this mix 1 one can of orange juice and 2 cans of water (1:2), students should recognize two things:

1. The ratio of orange juice to water in the combined mixture is no longer equivalent to the initial ratio but has changed to 4 cans of orange juice to 6 cans of water (4:6).
2. The ratio of orange juice to water in the combined mixture (4:6) is reduced from what it was initially (3:4).

In other words, students should be able to reason that the fractions $^3/_4$ and $^4/_6$ are associated with the ratios 3:4 and 4:6, respectively, and to deduce from the context that $^3/_4$ is greater than $^4/_6$ because the ratio of orange juice to water has been reduced. Furthermore, students should understand that although they would never write $^3/_4 + {^1/_2} = {^4/_6}$ to represent fractional ideas, the combining makes sense in a ratio context.

The same reasoning would allow a student to know that if initially 3 red and 5 blue marbles are in a box and 3 red and 4 blue marbles are added, the ratio of red marbles to blue marbles (or red marbles to total number of marbles) increases. Because the student understands that the ratio of red marbles to blue ones has increased, he or she should be able to compare the fractions associated with those ratios, $^3/_5$ and $^6/_9$ (or $^3/_8$ and $^6/_{15}$), recognizing that $^3/_5$ is less than $^6/_9$ (or $^3/_8$ is less than $^6/_{15}$). Giving students opportunities to think along these lines is a critical part of any instructional approach, since engaging in this kind of thinking allows them to broaden the scope of what it means to reason proportionally.

Knowledge of assessment

The use of formative assessment strategies, along with a good question or task, can help you to know and understand a great deal about your students' thinking.

Problem 3.1, which focuses on the equivalence of $^6/_9$ and $^{10}/_{15}$, was purposely chosen to address a misconception that arises when multiples in the numbers in the problem have no direct relationship. Sean showed evidence of this misconception when she indicated, erroneously, that no number exists to use as a multiplier of 6 for a product of 9 and as a multiplier of 10 for a product of 15 (see fig. 3.1). Lamon (2001) found that when given these numbers in a context, students were more likely to conclude that that they were equivalent. For example, students working with pies are more likely to change either 6 pies/9 pies and 10 pies/15 pies to $^{30}/_{45}$ and $^{30}/_{45}$ and know that they are equivalent. Thus, you should take care to choose a task or question and expect an explanation, whether the problem is situated in a context or not. The explanations will help you determine what your students know and see—not only whether their answers are correct or not. Their explanations can also tell you when you need more information from them to be certain of what they know. For instance, JQ claimed in his response to problem 3.1 that the fractions $^6/_9$ and $^{10}/_{15}$ are equivalent because you can multiply $^6/_9$ "by 5" and $^{10}/_{15}$ "by 3" and "they [are] equal" (see fig. 3.3). Without further elaboration, this response would not make it clear whether JQ, although he answered correctly, had grasped the idea of equivalence.

Conclusion

To help students develop a robust understanding of aspects of reasoning about ratios and proportions, this chapter has provided examples of student work and reflection questions to highlight ideas for working toward Essential Understanding 7. Wu (2009) notes, "Because fractions are a student's first serious excursion into abstraction, understanding fractions is the most critical step in understanding rational numbers and in preparing for algebra" (p. 8). Working through and elaborating on these examples and discussing them and others with students will help students understand—

- equivalent ratios and their creation by iterating or partitioning a composed unit;
- the relationships that are involved when one quantity in a ratio is multiplied or divided by a particular factor, including the proportional relationship that is maintained when the other quantity is multiplied or divided by the same factor; and
- the relationship that exists between multiplicative comparisons and composed units.

Constant consideration and discussion of the context in which any of these numbers arise is critical to students' development of deep understanding of these ideas.

into practice

Chapter 4
Reasoning about Rates

Essential Understanding 8

A rate is a set of infinitely many equivalent ratios.

Definitions of the terms *ratio* and *rate* in U.S. mathematics curricula often are not the same from one curriculum to another. Lobato and Ellis (2010) and Thompson (1994) present some of the ways in which these differences generally manifest themselves in curricula. Specifically, Thompson notes that the existence of the terms *ratio* and *rate* "would suggest that we have two ideas different enough to warrant different names. Yet, there is no conventional distinction between the two, and there is widespread confusion about such distinctions" (p. 189). Specifically, Thompson notes that three general distinctions are frequently made between ratio and rate:

1. A ratio compares quantities that are "like," whereas a rate compares quantities that are "unlike."
2. A ratio expresses the amount of one quantity with respect to another quantity, whereas a rate is a specific ratio between a quantity and an amount of time.
3. A ratio is a relation that can be expressed through ordered pairs of quantities, whereas a rate is one unit of a quantity as related to another quantity.

Because no one way of differentiating between *ratio* and *rate* is definitive, understanding ways in which students make sense of situations involving ratios used as rates is critical. In *Developing Essential Understanding of Ratios, Proportions, and Proportional Reasoning for Teaching Mathematics in Grades 6–8,* Lobato and Ellis

(2010) identify Thompson's (1994) conception of the idea of rate within the more general idea of ratio. For Thompson, the specific distinction of rate is that it signifies a set of infinitely many equivalent ratios. In other words, in this view, a rate is a constant ratio that can be reflectively abstracted to describe equivalent ratios.

Thompson's work suggests that teachers need to recognize the importance of interpreting rate in this way. To help students develop this perspective, teachers should engage them in building an understanding that does not rely on the situational context but rather enables them to delineate a finite set of equivalent ratios from a rate, with the set serving as an abstraction of the limited set of ratios in the context. Students with this orientation are likely to be reasoning about rate as a set of infinitely many equivalent ratios.

The focus of this chapter is on ways of helping students develop understandings of ratio and rate that are based on their own thinking and experience rather than on explicit or implicit referents found in typical problem situations. By supporting your students in recognizing their own conceptualizations of ratio and rate in problems, instead of dictating to them the construct that they should recognize as implied by the problem situation, you provide them with a critical foundation on which to build knowledge based on their conceptions.

Working toward Essential Understanding 8

The Common Core State Standards for Mathematics (CCSSM; National Governors Association Center for Best Practices and Council of Chief State School Officers [NGA Center and CCSSO] 2010) expect an instructional emphasis on ratio and rate concepts in grades 6 and 7. CCSSM indicates that grade 6 students should "understand ratio concepts and use ratio reasoning to solve problems" (p. 42), with a focus on using ratio and rate language in the context of ratio relationships to understand these relationships and conceptualize a unit rate associated with a ratio (see fig. 4.1).

CCSSM recommends that students in grade 7 "analyze proportional relationships and use them to solve real-world and mathematical problems" (p. 48). Specifically, with respect to understanding a rate as a reflectively abstracted constant ratio, CCSSM expects students to learn that ratios, in like *or* different units, have associated unit rates (constants of proportionality) that the students should be able to compute. Students are expected to identify and use this unit rate in various representations (tabular, graphical, symbolic, diagrammatic, and verbal), as well as in explaining what a point means on a graph that represents a proportional relationship (see fig. 4.2).

Ratios and Proportional Relationships	6.RP

Understand ratio concepts and use ratio reasoning to solve problems.

1. Understand the concept of a ratio and use ratio language to describe a ratio relationship between two quantities. *For example, "The ratio of wings to beaks in the bird house at the zoo was 2:1, because for every 2 wings there was 1 beak." "For every vote candidate A received, candidate C received nearly three votes."*
2. Understand the concept of a unit rate a/b associated with a ratio $a:b$ with $b \neq 0$, and use rate language in the context of a ratio relationship. *For example, "This recipe has a ratio of 3 cups of flour to 4 cups of sugar, so there is 3/4 cup of flour for each cup of sugar." "We paid \$75 for 15 hamburgers, which is a rate of \$5 per hamburger."*[1]
3. Use ratio and rate reasoning to solve real-world and mathematical problems, e.g., by reasoning about tables of equivalent ratios, tape diagrams, double number line diagrams, or equations.
   a. Make tables of equivalent ratios relating quantities with whole-number measurements, find missing values in the tables, and plot the pairs of values on the coordinate plane. Use tables to compare ratios.
   b. Solve unit rate problems including those involving unit pricing and constant speed. *For example, if it took 7 hours to mow 4 lawns, then at that rate, how many lawns could be mowed in 35 hours? At what rate were lawns being mowed?*
   c. Find a percent of a quantity as a rate per 100 (e.g., 30% of a quantity means 30/100 times the quantity); solve problems involving finding the whole, given a part and the percent.
   d. Use ratio reasoning to convert measurement units; manipulate and transform units appropriately when multiplying or dividing quantities.

Fig. 4.1. Grade 6 standards in the domain Ratios and Proportional Relationships, CCSSM 6.RP (NGA Center and CCSSO 2010, p. 42)

Identifying when your students are reasoning about ratios and when they are reasoning about rates is challenging, but discerning the difference is also crucial to recognizing the mathematical knowledge that underlies and supports their computations of unit rates and their use of rate language. To have a good understanding of students' conceptions of these constructs, particularly of a constant of proportionality, you first need to understand the ways in which your students are using ratio-oriented language and ideas, and the ways in which they are using rate-oriented language and ideas, in the case of a particular problem.

Ratios and Proportional Relationships **7.RP**

Analyze proportional relationships and use them to solve real-world and mathematical problems.

1. Compute unit rates associated with ratios of fractions, including ratios of lengths, areas and other quantities measured in like or different units. *For example, if a person walks 1/2 mile in each 1/4 hour, compute the unit rate as the complex fraction* $^{1/2}/_{1/4}$ *miles per hour, equivalently 2 miles per hour.*

2. Recognize and represent proportional relationships between quantities.

   a. Decide whether two quantities are in a proportional relationship, e.g., by testing for equivalent ratios in a table or graphing on a coordinate plane and observing whether the graph is a straight line through the origin.

   b. Identify the constant of proportionality (unit rate) in tables, graphs, equations, diagrams, and verbal descriptions of proportional relationships.

   c. Represent proportional relationships by equations. *For example, if total cost t is proportional to the number n of items purchased at a constant price p, the relationship between the total cost and the number of items can be expressed as t = pn.*

   d. Explain what a point *(x, y)* on the graph of a proportional relationship means in terms of the situation, with special attention to the points (0, 0) and (1, *r*) where *r* is the unit rate.

3. Use proportional relationships to solve multistep ratio and percent problems. *Examples: simple interest, tax, markups and markdowns, gratuities and commissions, fees, percent increase and decrease, percent error.*

Fig. 4.2. Grade 7 standards in the domain Ratio and Proportion, CCSSM 7.RP (NGA Center and CCSSO 2010, p. 48)

Problem 4.1 presents a task that requires ratio- or rate-oriented thinking:

Problem 4.1

To make Luscious Lilikoi Punch, Austin mixes $^1/_2$ cup lilikoi passion fruit concentrate with $^2/_3$ cups water. If he wants to mix concentrate and water in the same ratio to make 28 cups of Luscious Lilikoi Punch, how many cups of lilikoi passion fruit concentrate and how many cups of water will Austin need?

Reflect 4.1 prompts you to think about ways in which you—and your students—might address this problem by using ratio-oriented language, as well as how you—and they—might address it by using rate-oriented language.

Reflect 4.1

Consider problem 4.1. What approach could you take to indicate that you are conceiving of the situation as involving ratios—that is, that you are comparing two quantities multiplicatively?

What approach could you take to indicate that you are conceiving of the situation as a rate—that is, that you are abstracting a constant ratio to describe a rate?

Problem 4.1 presents students with an initial ratio of fruit concentrate to water, both measured in units of cups. Austin has found, presumably through many experiments and recipe refinements, that he likes his Luscious Lilikoi Punch best when the ratio of cups of concentrate to cups of water is $^1/_2 : {}^2/_3$. Most mathematics curricula would present this problem as a ratio situation simply because it gives both of the quantities in the ratio in the same units.

However, this argument might not necessarily be the best one to support interpreting this scenario as a ratio situation. What might be a better argument for considering the situation from the standpoint of ratios? Beyond noting the "same units versus different units" distinction between ratio and rate contexts, respectively, what interpretation of this problem would indicate a conceptualizing of it from a ratio-as-multiplicative-comparison perspective?

In examining the problem, you might quickly focus on the fact that 1 mixture of concentrate and water gives Austin $^7/_6$ cups, or 1 and $^1/_6$ cups, of punch. If you realize that by making a certain number of mixtures, Austin will make a full cup out of the fractional $^1/_6$ cup that results in 1 mixture, you can reason that by making 6 mixtures, Austin will produce 6 and $^6/_6$ cups, or 7 cups of punch. In this way, you have effectively determined a scalar to relate the ratio of $1 : {}^7/_6$ to 6 : 7 mixtures to cups of punch. With this scale factor of 6, you then know that 6 mixtures will require 3 cups of concentrate ($6 \times {}^1/_2$ cups concentrate) and 4 cups of water ($6 \times {}^2/_3$ cups water). Finally, to determine how many mixtures Austin will need for 28 cups of punch, you know that he will need four times as many cups as he can make with 6 mixtures—7 cups (3 cups concentrate plus 4 cups water). In other words, you can form a new ratio, 24 : 28, of mixtures to cups of punch, by comparing quantities multiplicatively, using the scalar factor of 4. Thus, if 6 mixtures will give Austin 7 cups of punch with 3 cups of concentrate and 4 cups of water, then 24 mixtures will give him four times as many cups, or 28 cups, of punch, with 12 cups of concentrate and 16 cups of water.

Although this is only one possible interpretation, it illustrates how you, or your students, could make sense of this problem from one perspective—that is, by attending to various ratios and multiplicative comparisons between quantities. How, by contrast, might you, or they, conceptualize this problem, which seemingly involves *same* units, from another perspective—that is, by focusing on a rate as an abstracted ratio?

Although you know how many cups of punch 1 mixture of the ingredients makes ($^7/_6$), you do not immediately know how many cups of each ingredient Austin would need to produce 1 cup of punch. However, if you begin by reasoning from a ratio perspective, you know that 3 cups of concentrate and 4 cups of water will produce 7 cups of punch. This reasoning is similar to the thinking explored previously. But if you use the ratios 3:7 and 4:7 to understand a *unit rate* in the way that CCSSM expects, your reasoning will then take you in a different direction. You know that a ratio of 3 cups of concentrate to 7 cups of punch means that Austin will have $^3/_7$ cups of concentrate for every cup of punch. Similarly, you know that for every cup of punch there are $^4/_7$ cups of water. From these relationships—from this rate—you can abstract any ratio of cups of one ingredient for cups of punch.

To examine these ratio and rate perspectives further, suppose, for example, that you want to know how many cups of concentrate are in $2^1/_3$ cups of punch left after a party. You can approach this situation by knowing that there are 3 cups of concentrate for every 7 cups of punch. You might then recognize that the amount of punch left—$2^1/_3$ cups—is one-third of the amount of punch in that unit ratio. You could then rely on your understanding of invariance in ratio relationships to find one-third of the cups of concentrate in your initial ratio, and thus, you could determine that there is 1 cup of concentrate in the punch that is left at the party.

Alternatively, you could apply your knowledge of the unit rate, reasoning from the fact that there are $^3/_7$ cups of concentrate for every cup of punch. Hence, if you are determining the amount of concentrate in $2^1/_3$ cups of punch left after a party, then you know that this amount is two $^3/_7$ cups of concentrate plus one-third $^3/_7$ cups of concentrate, or $(2 \times {}^3/_7) + ({}^1/_3 \times {}^3/_7) = ({}^6/_7 + {}^1/_7) = 1$ cup of concentrate.

On one hand, by thinking about this problem from a ratio-as-multiplicative-comparison perspective, students can approach it with scalars—a concept with which they may have enough experience to feel fairly comfortable. On the other hand, by thinking about the problem from the perspective of a rate as an abstraction of the ratios involved, they can arrive at a unit rate that they can use to answer questions that scaling perspectives do not necessarily let them approach easily. As Lobato and Ellis (2010) argue, “Thinking about rate in this manner can help you guard against overestimating your students’ proportional reasoning abilities”

(p. 43). For instance, using a rate of ${}^{3}/_{7}$ per 1 allows students to form ratios beyond 3:7. They should be able to identify ratios such as ${}^{6}/_{7}$ per 2 and ${}^{18}/_{7}$ per 6, or ${}^{12}/_{21}$ per ${}^{4}/_{3}$ and ${}^{27}/_{14}$ per ${}^{9}/_{2}$.

These examples are but a few of the many that you and your students will encounter as you work together to develop their understanding as expected in CCSSM for grades 6 and 7. However, the examples illustrate and underscore the important perspectives that your students must take when analyzing and solving problems—and that you must take when examining and assessing their work and making evaluative judgments regarding their mathematical abilities to work with ratios and proportions and use proportional reasoning.

Figures 4.3 and 4.4 present the work of two seventh-grade students, Grayson and Eileen, on problem 4.1. Use the question in Reflect 4.2 to guide you in considering what their solutions might indicate about their approaches to the problem.

Reflect 4.2

Inspect the work on problem 4.1 by Grayson and Eileen, presented in figures 4.3 and 4.4, respectively. As evidenced by their representations, explanations, and strategies, how do you think that these students were conceptualizing the situation?

Were they reasoning about ratios through multiplicative comparisons or reasoning about rates through abstracted ratios?

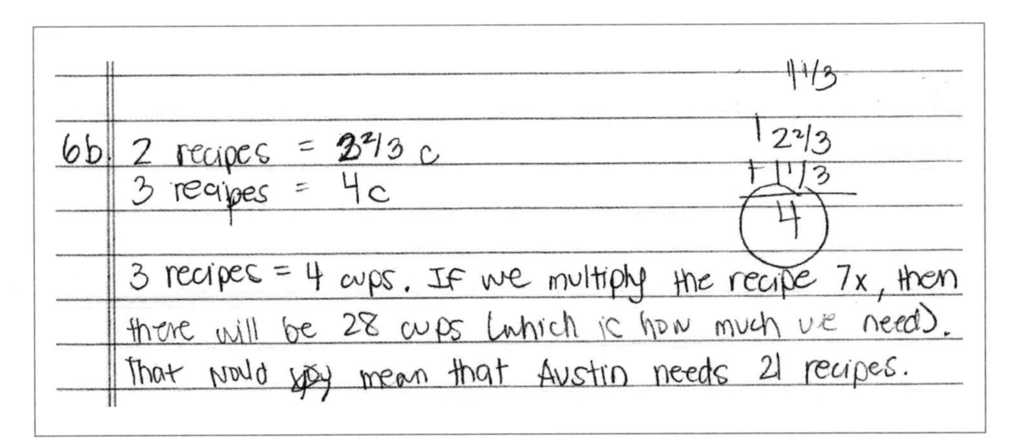

Fig. 4.3. Grayson's work on problem 4.1

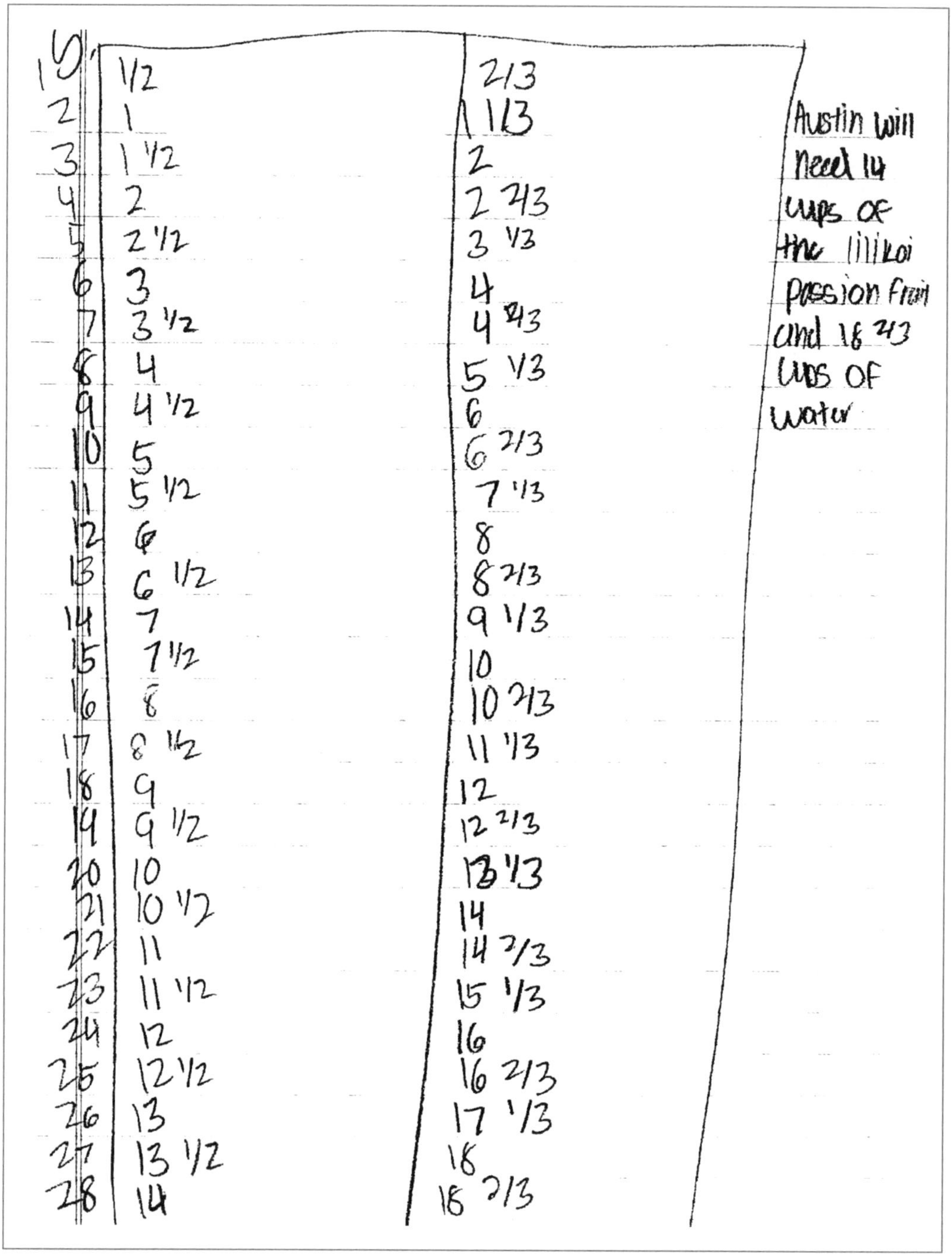

1	1/2	2/3
2	1	1 1/3
3	1 1/2	2
4	2	2 2/3
5	2 1/2	3 1/3
6	3	4
7	3 1/2	4 2/3
8	4	5 1/3
9	4 1/2	6
10	5	6 2/3
11	5 1/2	7 1/3
12	6	8
13	6 1/2	8 2/3
14	7	9 1/3
15	7 1/2	10
16	8	10 2/3
17	8 1/2	11 1/3
18	9	12
19	9 1/2	12 2/3
20	10	13 1/3
21	10 1/2	14
22	11	14 2/3
23	11 1/2	15 1/3
24	12	16
25	12 1/2	16 2/3
26	13	17 1/3
27	13 1/2	18
28	14	18 2/3

Austin will need 14 cups of the lilikoi passion frui and 18 2/3 cups of water

Fig. 4.4. Eileen's work on problem 4.1

As illustrated by his work in figure 4.3, Grayson approached this problem by looking at "recipes"—that is, mixtures of the ingredients called for in the recipe in the quantities specified. He first stated that 2 recipes are equal to $2^2/_3$ cups, presumably of punch. Working with this ratio, he added $1^1/_3$ (cups) to the amount of punch to get a ratio of 3 recipes to 4 cups of punch.

Perhaps an initial reaction to Grayson's work is that it is simply incorrect. Grayson made an error in his result for 2 recipes, obtaining $2^2/_3$ instead of $2^1/_3$ cups of punch, and he appears to have built on that result to obtain his result for 3 recipes: $2^2/_3 + 1^1/_3 = 4$ cups. However, to understand the perspective from which Grayson was approaching the problem and develop his mathematical abilities by building on what he was able to demonstrate, his teacher would do well not to dwell on the incorrectness of his solution but rather to recognize the important mathematical capacities that Grayson was displaying in his work.

Such analysis of student work allows teachers to scaffold questions according to what their students have displayed of their thought processes. Specifically, in this case, Grayson moved from the ratio 2 recipes to $2^2/_3$ cups of punch to the ratio of 3 recipes to 4 cups of punch. In these two lines of his solution, he provided insight into ideas that he held and assumptions that he made that helped him understand the problem. In particular, because he gave a ratio for 2 recipes, and another ratio for 3 recipes, his teacher might suspect that he thought 1 recipe produces $1^1/_3$ cups of punch. By analyzing the student's work in this way, the teacher could narrow the possibilities to determine the misconception that kept Grayson from moving through the problem successfully. Was he held back by an arithmetic miscalculation of $^1/_2 + ^2/_3$, or did he perhaps add correctly and obtain $^7/_6$ but mishandle the doubling process for 2 recipes? Or did he misinterpret the problem? The next step for the teacher after determining the nature of Grayson's incorrect answer would be to identify the approach to reasoning about ratios and rates that the student was taking. Given that he stopped with the ratio 3:4 and correctly scaled that ratio by 7 to say that 21 recipes were needed to make 28 cups of punch, we can postulate that Grayson was likely to have been reasoning about ratios as multiplicative comparisons.

Eileen approached the problem by making a table, as shown in figure 4.4, listing the cups of each ingredient necessary to make increasing numbers of mixtures. However, Eileen seems to have misinterpreted her representation, because she ultimately determined the solution on the basis of 28 mixtures rather than 28 cups of mixture. In fact, examination of Eileen's work reveals that she correctly found the amount of each ingredient needed for 24 mixtures—12 cups of concentrate and 16 cups of water, for 28 cups of punch—the total number of cups desired in the problem situation—and $^6/_7$ of the amounts that Eileen gave for 28 mixtures. Consequently, an initial question from the teacher might investigate how Eileen was conceiving of the relationship between total numbers of mixtures and total numbers of cups of punch.

What might appear to be most significant in Eileen's work is her use of a table that presents equivalent ratios of cups of concentrate to cups of water in growing numbers of mixtures of punch. Categorizing her thinking as reflecting an approach to rate as an abstracted constant ratio is tempting—a constant rate through which any of the infinitely equivalent ratios that exist in the proportional relationship can be determined. However, further reflection on Eileen's work and scrutiny of it underscore the need for the caution urged by Lobato and Ellis (2010) against overestimating students' proportional reasoning abilities. In other words, Eileen's ability to create a table of equivalent ratios does not guarantee that she was approaching this problem from a rate-as-an-abstracted-ratio perspective. Even though Eileen successfully created the table, her work did not give any indication that she used the constant ratio of $^1/_2$ cup concentrate to 1 mixture, or $^2/_3$ cups water to 1 mixture, to find any non-whole number quantities of mixtures to gain more understanding or solve the problem.

As a result, Eileen's response to problem 4.1 presents an important illustration. Her work seems to suggest that she was conceiving of the problem in terms of a rate, particularly since she presented a table that many adults and teachers would look at as a representation of a constant rate of change per mixture. However, because Eileen did not use any ratio in a more abstracted manner, it seems likely that she was approaching this problem simply through an iteration of the given ratios and presenting her results in table form.

Summarizing Pedagogical Content Knowledge to Support Essential Understanding 8

Teaching the mathematical ideas in this chapter requires a specialized knowledge related to the four components presented in the Introduction: learners, curriculum, instructional strategies, and assessment. The four sections that follow summarize some examples of these specialized knowledge bases in relation to Essential Understanding 8. Although we separate them to highlight their importance, we also recognize that they are connected and support one another.

Knowledge of learners

In *Developing Essential Understanding of Ratios, Proportions, and Proportional Reasoning for Teaching Mathematics in Grades 6–8*, Lobato and Ellis (2010) note, "Thompson argues that these [curricular] definitions [of rate and ratio] locate the distinction between ratio and rate in the *situation* rather than in the way that a student *conceives* of the situation" (p. 42). In other words, the various ways in which rates and ratios are generally defined in mathematics curricula focus on characteristics of the situations (same units versus different units) rather than how students are working with the situations and interpreting them (perhaps in terms of "recipes" or iterating unit mixtures through a table that mirrors some process of generating equivalent ratios that may confound teachers' abilities to quickly identify conceptions of rate in students' work).

When Thompson (1994) discusses the ways in which he delineates interpreting a problem through multiplicative comparisons of ratios versus interpreting it by abstracting the ratios to determine a rate, he notes, "A rate, as a reflectively abstracted constant ratio, symbolizes that structure as a whole, but gives prominence to the constancy of the result of the multiplicative comparison" (p. 192). In other words, in examining the work of Grayson and Eileen (see figs. 4.3 and 4.4, respectively), you might say that neither student necessarily gave prominence to the constancy of the result of any multiplicative comparisons. On one hand, in considering "recipes" Grayson focused on using multiplicative comparisons to arrive at a meaningful statement about how many recipes Austin will need to make 28 cups of punch. On the other hand, although Eileen presented her work through a table of the sort that often can be related to some sort of rate-oriented thinking, she did not appear to give prominence to the constancy of a multiplicative comparison. In fact, whether Eileen approached the problem multiplicatively at all is debatable. It seems more likely that she approached it additively. However, such speculations illustrate why it is important to approach students' understandings from *their* perspectives, and doing so requires further understanding of where these problems occur in the curriculum, as well as knowledge of the ways in which students' work on the problems is assessed—and effective strategies for probing students' thinking.

Knowledge of curriculum

An integral aspect of your knowledge of curriculum is your ability to select tasks that will allow your students to work from their own understanding to develop solution strategies that you can then use to engage them in robust discussions. For the most part, such tasks do not arise naturally in classroom discussions but depend on your recognition of the need to offer them to challenge your students and your willingness to allow time for student discussions of them. As quoted in Chapter 1, Steffe (1994) observed, "The current notion of school mathematics is based almost exclusively on formal mathematical procedures and concepts that, of their nature, are very remote from the conceptual world of the children who are to learn them" (pp. 5–6). In relation to Essential Understanding 8, the question for you becomes, "What is the conceptual world of students who are to understand concepts related to ratio and rate in grades 6–8?"

Thompson (1994) again provides some frame of reference for answering this question. In discussing ways of understanding students' conceptions of ratios as rates, he offers the following observations (p. 193):

> The first sense of reflected abstraction is that a class of actions are reconstructed and symbolized at the level of mental operations. The second sense of reflected abstraction is that as a situational conception is constructed, the figurative aspects of the conception are "reflected" to the level of mental operations. In the first sense, we would say a person has learned (e.g., learned speed as a rate [or, in the case of problem 4.1, mixture as a rate]). In the second sense, we would say a person has comprehended (e.g., conceived an object's motion as a rate [or, in problem 4.1 the flavor—*lilikoi-ness*—of the punch]).

In other words, it is worth considering students' conceptions of rate problems even beyond notions of ratio. That is, students may conceive of problem 4.1 at the level of different mixture rates to make larger or smaller amounts of punch. However, students should be encouraged to analyze and reconceive of the punch problem from a different standpoint, by considering that any quantity of punch should have the same flavor as any other quantity. That is, the constancy of "lilikoi-ness" in the punch should be evident in any ratio that describes the situation.

Consequently, attending to students' thinking when they are engaging in discussions and presenting solutions to ratio problems is critical. If your students are conceiving of situations through multiplicative comparisons, you should identify follow-up questions or tasks that you can provide to challenge them to think about the constancy that they have found in terms of other equivalent ratios. These questions or tasks will prompt students to define a rate that is related to the initial ratio. In attending to your students' thinking in this way, you are attempting to give it coherence that can extend your students' knowledge base for future work.

Knowledge of instructional strategies

This chapter has provided samples of instructional strategies for developing students' mathematical understanding by focusing on their thinking and strategies. If your students respond to a task like problem 4.1 by using only strategies similar to those of Grayson and Eileen, you must make an instructional decision about how to bring other approaches into general discussions. Those strategies could mirror constructs from the discussion following Reflect 4.1. Recall also that Chapter 1 drew attention to Lamberg's (2012) strategies for facilitating "math talk" in peer, small-group, and whole-class discussions, generating other strategies that you can highlight as you facilitate the conversations.

The characteristics that Lamberg (2012) observes in discussions that facilitate the development of students' understanding are worth restating:

- Students develop "shared" understanding of a problem.
- Teachers provide students with "guided instruction."
- Students evaluate and analyze their thinking and their peers' thinking.

Likewise, Lamberg's three phases (p. 11) that teachers should go through in the course of facilitating mathematical connections in whole-class discussions are useful to reiterate:

Phase 1: Making thinking explicit

Phase 2: Analyzing solutions

Phase 3: Developing new mathematical insights

The teacher facilitates the discussion through questioning.

Similarly, the value of Smith and Stein's (2011) five practices for productive discussions, also discussed in Chapter 1, cannot be overstated. Remember, this sequence of practices can enable you to orchestrate mathematical discussions through–

- anticipating students' solutions;
- monitoring students' in-class work;
- selecting approaches for students to share;
- sequencing those approaches purposefully to maximize impact and learning; and
- connecting the approaches and the underlying mathematics.

Making students' thinking explicit to the students themselves as well as their classmates has tangible benefits. By connecting the ways in which ratios have associated rates that can be used to describe and model situations that go beyond equivalent ratios found through scaling "easy" ratios, you highlight for your students the importance of understanding and being aware of their own conceptions of situations and the mathematical significance of those conceptions.

Knowledge of assessment

Wiliam (2007) highlights the importance of formative assessments–assessments that allow students to share their thinking and teachers to make prompt instructional decisions based on the understanding or misunderstanding that they demonstrate. The samples of work from Grayson and Eileen provided insight into their approaches to a problem largely involving ratios with similar units, although with a confounding unit of *mixture* as well. Such insight was possible only because the students were asked to show their thinking on a task that, as recommended in the Introduction, encourages reversibility, flexibility, and generalization in students' thinking (Dougherty 2001).

Considering assessment possibilities is a critical part of determining tasks to give students, since rich possibilities for follow-up questions are essential for further assessment of student thinking. With students such as Grayson and Eileen, a teacher who knows that students' conceptions of a situation largely determine their strategies related to ratio or rate will be able to ask follow-up questions to clarify their reasoning. As discussed in the previous section, these follow-up questions can be addressed to students individually or in peer, small-group, or whole-class discussions. If a student is examining a problem from one perspective, the teacher should prompt him or her to rethink the strategy in use by asking questions such as, "Can you talk to me or your partner(s) about other quantities that you can make from the relationships you see in this problem?"

Conclusion

The discussion and students' work provided in this chapter have illustrated mathematics that you need to understand so that you can understand your students as individual learners, design opportunities for them to learn, make critical instructional decisions, and meaningfully understand students' responses to assessments. As a teacher, you need to understand ways in which your students are conceiving ratio problems. Through understanding your *students'* perspectives, you can more effectively engage them in thinking deeply, robustly, and accurately about concepts related to ratio and rate.

into practice

Chapter 5 Generalizing Reasoning to Solve Proportion Problems

Essential Understanding 9

Several ways of reasoning, all grounded in sense making, can be generalized into algorithms for solving proportion problems.

For students to develop their understanding of proportions and their skill in solving proportion problems, they need to explore proportional relationships in many settings and contexts and generalize their reasoning. This chapter provides examples of problems and situations that encourage students to conceptualize the relationships in ways that can lead them to make generalizations and build coherent representations that are related to "rules" for solving proportion problems. In particular, the examples in the chapter, some of which appeared earlier in the book, offer situations that support sense making that can lead to two algorithms—the means-extremes product property for proportions and the process for dividing fractions by multiplying by the reciprocal. The means-extremes product property for proportions is stated in various ways, all expressing the same idea: in a proportion, the product of the means equals the product of the extremes. This process is sometimes referred to simply as "cross multiplication," and examining it can help students understand relationships between numbers and their reciprocals—in particular, if $a < b$, then the reciprocals of a and b, $1/a$ and $1/b$, are related as $1/a > 1/b$.

Working toward Essential Understanding 9

"Reason abstractly and quantitatively" is standard 2 in the Standards for Mathematical Practice in the Common Core State Standards for Mathematics (CCSSM; National Governors Association Center for Best Practices and Council of Chief

State School Officers [NGA Center and CCSSO] 2010). As noted in Chapter 2, CCSSM (p. 6) describes and expands on the complementary abilities involved in the essential practice identified in this standard:

> Mathematically proficient students ... bring two complementary abilities to bear on problems involving quantitative relationships: the ability to *decontextualize*—to abstract a given situation and represent it symbolically and manipulate the representing symbols as if they have a life of their own, without necessarily attending to their referents—and the ability to *contextualize*, to pause as needed during the manipulation process in order to probe into the referents for the symbols involved.

Although mathematics teachers and curricula often make efforts to situate examples in context so that students have background knowledge from which to work, this statement emphasizes the importance of prompting students to move in the opposite direction as well. That is, one aspect of reasoning abstractly and quantitatively comes into play when students try to solve problems that present numbers without a context. In such cases, students should be encouraged to try to contextualize what is abstract—to try to envision a context with which they might associate the numbers and then work with that context if doing so is helpful in answering the question. In addition to having your students think of a context for the numbers, you should also ask them to compare and contrast the ideas and contexts generated by students.

Problem 5.1 asks students to compare two numerical values that are not equal. This task provides them with an opportunity to consider a context for the numbers and use it to solve the problem.

> **Problem 5.1**
>
> As much as possible, use only mental arithmetic to determine which is larger, $^{14}/_{29}$ or $^{15}/_{31}$.

Reflect 5.1 gives you a chance to think about contexts that you might use to solve the problem.

Reflect 5.1

In deciding which of the two numbers in problem 5.1 is larger, what ideas did you consider to determine your response mentally, without pencil and paper?

Did your thinking involve fractions representing customary part-whole relationships or did it involve ratios?

Students' responses to problem 5.1 offer opportunities to explore their reasoning and compare and contrast thinking that involves fractions representing part-whole relationships and thinking that involves ratios. The responses shown in figures 5.1–5.5 were collected from five middle-grades students who were given this problem. Note that the problem is stated without a context, so some of the responses feature thinking based on fractions, and some feature thinking based on ratios. You might use such samples in the classroom, comparing and contrasting the thinking to strengthen your students' understanding of ratios and to demonstrate how thinking in terms of ratios can also help strengthen knowledge of fractions.

$\frac{14}{29}$ is larger because cutting a pie into 29 pieces gives larger pieces than if cut into 31 pieces.

Fig. 5.1. Aaron's response to problem 5.1

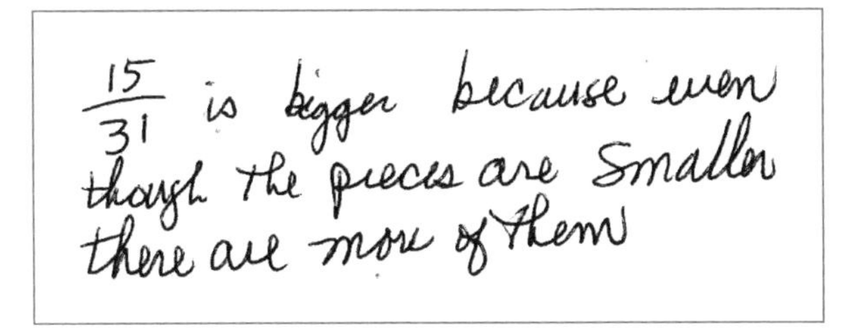

Fig. 5.2. Laverne's response to problem 5.1

$$\frac{14}{29} \cdot \frac{31}{31} = \frac{434}{29 \cdot 31}$$

14
31
14
42

$$\frac{15}{31} \cdot \frac{29}{29} = \frac{435}{29 \cdot 31}$$

15
29
135
30

Fig. 5.3. Atsuko's response to problem 5.1

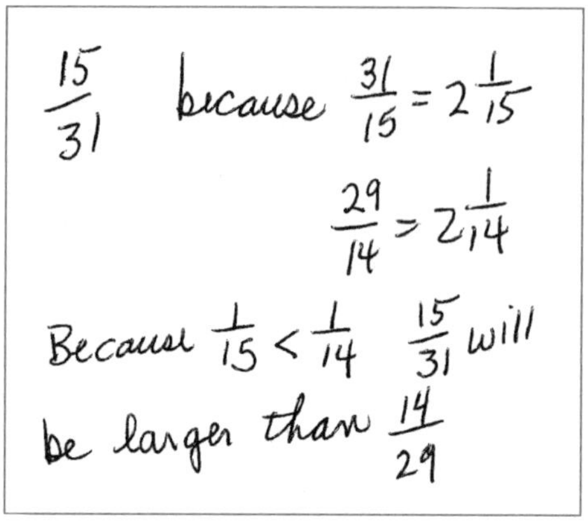

$\frac{15}{31}$ because $\frac{31}{15} = 2\frac{1}{15}$

$\frac{29}{14} = 2\frac{1}{14}$

Because $\frac{1}{15} < \frac{1}{14}$ $\frac{15}{31}$ will be larger than $\frac{14}{29}$

Fig. 5.4. Scott's response to problem 5.1

$\frac{15}{31}$ becquse $\frac{14}{29} + \frac{1}{2} = \frac{15}{31}$

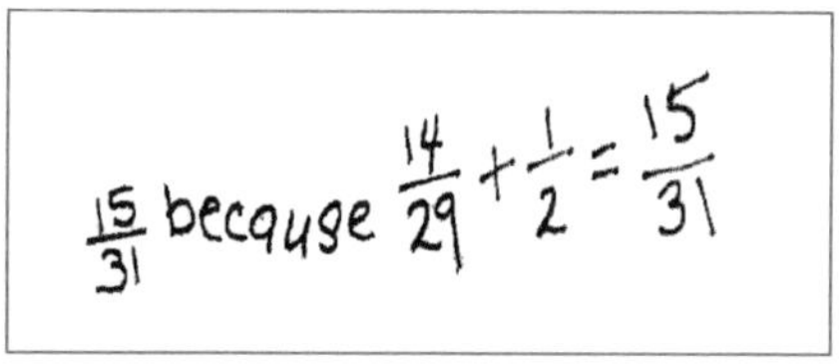

Fig. 5.5. Adele's response to problem 5.1

Although Aaron and Laverne, whose work appears in figures 5.1 and 5.2, respectively, focused on important considerations of fractions, they did not attend to the relationships between *both* the size of the pieces *and* the number of pieces. To work successfully with ratios and proportions, students must attend to multiple relationships at the same time. Both Aaron and Laverne recognized that partitioning a unit into 29 equal pieces makes pieces that are bigger than the pieces made by partitioning the same unit into 31 equal pieces. Although both students seemed to recognize that a smaller denominator yields a larger piece, they focused only on one comparison related to each number—either the number of pieces (14 versus 15) or size of the pieces ($^1/_{29}$ versus $^1/_{31}$)—without exploring the relationship between these comparisons.

Reasoning based on a single comparison related to the numbers in a problem is not uncommon in middle school students who are beginning to work with ratios and proportions. Chapter 2 also showed samples of work from students whose reasoning demonstrated this limitation. Recall the seventh-grade students' reasoning about the gumball setting in problem 2.2. In making comparisons to decide whether the chance of drawing a red gumball was greater from bag A with 10 red and 15 white gumballs or from bag B with 6 red and 8 white gumballs, Moses focused only on the difference between the number of red and white gumballs in each bag—5 for bag A and 2 for bag B. Although Kilani tried several ideas, her final explanation, "It is better have a less diff.," shows that her thinking also involved only one comparison, which in the case of this problem is not sufficient.

Atsuko, whose response is shown in figure 5.3, clearly did more than mental arithmetic to make the comparison. Atsuko saw that it would be useful to find a common denominator, and in the process she multiplied the pairs of numbers 14 and 31, and 15 and 29. Although Atsuko does not explicitly say so, her work implies that she recognized that $^{15}/_{31}$ is larger after she changed the fractions $^{14}/_{29}$ and $^{15}/_{31}$ to equivalent fractions with the same denominator, 29×31. It is worth noting that Atsuko saw no need to compute that product but did determine the products 14×31 and 15×29, from which she seems to have concluded that $^{15}/_{31}$ is larger.

Atsuko's work indicates relational reasoning. Although she was not necessarily aware of it, her reasoning leads to the means-extremes product property, often used in working with proportion problems. In fact, Atsuko's effort to find a common denominator could be used as a justification of this property because in a proportional setting,

$$\text{if} \quad \frac{a}{b} = \frac{c}{d}, \quad \text{then} \quad \frac{a \times d}{b \times d} = \frac{c \times b}{d \times b}, \quad \text{and hence,} \quad a \times d = c \times b.$$

Furthermore, it does not matter whether the thinking that results in this statement is in terms of fractions or ratios; the statement is true regardless.

Moreover, although such reasoning is often used to determine or demonstrate whether two ratios are equivalent or to solve problems with an unknown as one of four terms in a proportion, Atsuko's work demonstrates that the property can also be applied to determine which of two ratios is larger or smaller. Her work shows that to determine which ratio (or fraction) is larger, $^{14}/_{29}$ or $^{15}/_{31}$, she could apply the means-extremes product property to obtain 14×31 and 15×29. Because $14 \times 31 = 434$ and $15 \times 29 = 435$, she could see that $^{15}/_{31}$ is larger. Students can use Atusko's sense making to compare any two ratios, finding out not only whether they are equivalent but also which is larger if they are not equivalent.

Scott, whose work is shown in figure 5.4, saw that problem 5.1 could be reduced to an argument similar to Aaron's in figure 5.1. Unlike Aaron's solution, Scott's response centers on two mixed fractions that Scott created from the reciprocals of the given numbers and then compared by focusing on the unit fraction portion of the fractions. It is significant that Scott used reciprocal reasoning–which could also be useful in understanding the gumball setting in problem 2.2 in Chapter 2 and the relationship between units when looking at equivalent fractions on a number line in Chapter 3 (see fig. 3.7 and related discussion). That is, Scott reasoned that if $a < b$, then the reciprocals $1/a$ and $1/b$ are related by $1/a > 1/b$.

This property of reciprocals is related to the means-extremes product property. If a and b are positive numbers with $a < b$, you can rewrite $a < b$ as $a \times 1 < b \times 1$. If a and its associated 1 are considered as the extremes and b and its associated 1 are considered as the means, you can write

$$\frac{1}{b} < \frac{1}{a}.$$

Alternatively, multiplying both sides of the original inequality, $a < b$, by

$$\frac{1}{a \times b}$$

gives an algebraic proof of the same relationship.

Consider how you might use this reasoning in a proportional setting. Suppose that you have two containers, A with 4 red balls and 9 blue ones and B with 7 red balls and 11 blue ones. If asked which has a larger ratio of red to blue balls, you would choose B because 4:9 < 7:11 (or $4/9 < 7/11$). Furthermore, knowing that B has the larger ratio of red to blue balls would immediately tell you that A has a larger ratio of blue to red balls, even without considering the numbers involved. Yet, if someone asked you directly which had a larger ratio of blue to red balls, you would ordinarily look at the ratios 9:4 and 11:7 (or $9/4$ and $11/7$) and choose A. As this example suggests, it is natural to use reciprocals of ratios in many cases when considering comparisons of ratios.

Likewise, students need to understand that if they know the relationship between two quantities expressed as ratios (or fractions), they know the relationship between their reciprocals as well. This explains the importance of the reasoning demonstrated by Scott, who encountered a situation that perhaps was not clear to him, but by using reciprocals, he was able to reframe it in a context in which he was able to work and reason. In doing so, Scott demonstrated an ability to "reason abstractly and quantitatively," practice 2 in CCSSM's Standards for Mathematical Practice (NGA Center and CCSSO 2010, p. 6). By applying Scott's method to problem 3.1 about whether $6/9$ and $10/15$ are equivalent, students could consider $9/6$ and $15/10$. Because each of these is equivalent to 1.5, they would know that the two fractions are equivalent.

Adele's response appears in figure 5.5. When Adele explained her work orally, she revealed that she had done some interesting arithmetic that incorporated several principles of proportional thinking. Also engaging in practice 2, "Reason abstractly and quantitatively" (NGA Center and CCSSO 2010, p. 6), Adele contextualized this problem into one that made sense for her. In the words used in CCSSM to elaborate on practice 2, Adele was able "to pause as needed during the manipulation process in order to probe into the referents for the symbols involved" (p. 6). In verbalizing her thinking aloud, Adele said,

> I thought of this like batting averages in baseball. Before I played a game, I had 14 hits in 29 at-bats. During the game, I got 1 hit in 2 at-bats. My new batting average is 15 hits in 31 at-bats. Because I started the game hitting less than .500 but hit .500 during the game, my batting average went up.

Although the representation $^{14}/_{29} + {^1}/_2 = {^{15}}/_{31}$ is technically not correct, it makes sense in the context given orally by Adele because of the actions involved in the setting. This line of reasoning is applicable in several settings. For example, if a mixture that contains 14 cups of orange juice to 29 cups of water (14 : 29) has one cup of orange juice and two cups of water (1 : 2) added to it, the resulting mixture (15:31) will have a stronger concentration of orange juice than the original. So too, it makes sense that the $^{15}/_{31}$ batting average is greater than the $^{14}/_{29}$ average. These examples show that appropriate reasoning about ratios and proportional reasoning is also helpful in understanding relationships with fractions.

A return to problem 3.2 and students' reasoning about it provides an opportunity to see how students' responses can lead them to generalized processes for solving proportion problems and can also provide a justification for a common procedure for dividing fractions. Problem 3.2 is presented again below, here as problem 5.2:

> **Problem 5.2**
>
> Jonnine had a board. She cut and used $^2/_5$ of the board for bracing. She measured the piece used for bracing and found it to be $^3/_4$ foot long. How long was the original board? (Presented in chapter 3 as problem 3.2.)

Figures 5.6–5.8 show the reasoning of three students in grades 5 and 6 on this problem. Use the question in Reflect 5.2 to guide your examination of their thinking processes.

Reflect 5.2

Figures 5.6, 5.7, and 5.8, show, respectively, the responses of Emma, Kirk, and Rochelle to the board setting presented in problem 5.2. What reasoning and sense making do these students exhibit that might support generalizations that they could use in other settings?

2/5 = 3/4
4/5 = 6/4
5/5 = 7.5/4
the board was a foot and 7/8 long

Fig. 5.6. Emma's response to problem 5.2

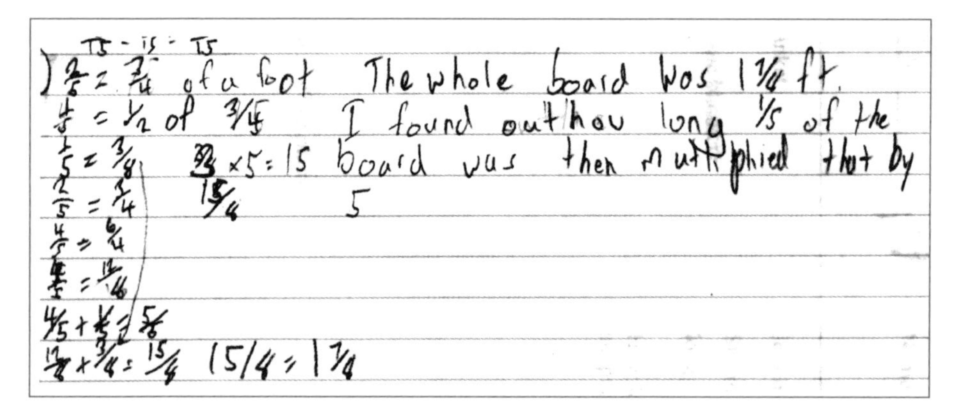

Fig. 5.7. Kirk's response to problem 5.2

2/5 = 3/4
x 2 1/2
5/5 = 15/8
5/5 = 1 7/8 ft.
The original board was 1 7/8 feet long.

Fig. 5.8. Rochelle's response to problem 5.2

Although Emma's work is written in a manner that is not symbolically correct ($^2/_5 = {}^3/_4$, and so on), Emma demonstrated proportional reasoning as she built from

$^{2}/_{5}$ of the board to $^{4}/_{5}$ of the board, showing that she understood the equivalence of the proportions that she was building:

$$\frac{2}{5}:\frac{3}{4}=\frac{4}{5}:\frac{6}{4}$$

Yet Emma did not clarify how she went from $^{4}/_{5}$ of the board to $^{5}/_{5}$ of the board. She also used an interesting fraction, $^{7.5}/_{4}$, in her solution, although she changed it to a mixed number, $1^{7}/_{8}$. Whether Emma conceived of her work additively or multiplicatively, she could have used the same reasoning to facilitate work in an algebraic setting. For example, if you know that

$$\frac{2}{5}x=\frac{3}{4}y,$$

then you can also correctly write

$$\frac{4}{5}x=\frac{6}{4}y.$$

Kirk and Rochelle extended this thinking in their responses.

Kirk's work, shown in figure 5.2, illustrates several important elements of proportional reasoning. It has components of the additive building-up process seen in Emma's work, as demonstrated by Kirk's combining $^{4}/_{5}$ of the board and $^{1}/_{5}$ of the board to get $^{5}/_{5}$ of the board and concomitantly combining $^{12}/_{8}$ feet and $^{3}/_{8}$ feet to obtain $^{15}/_{8}$ feet as the measurement associated with $^{5}/_{5}$ of the board. Although Kirk used the same sort of technically incorrect notation that Emma used, he demonstrated an understanding of proportionality by moving from "$^{2}/_{5} = {}^{3}/_{4}$ of a foot" to "$^{1}/_{5} = {}^{3}/_{8}$ of a foot." Kirk's notations shows that he understood the equivalence of these proportions:

$$\frac{2}{5}:\frac{3}{4}=\frac{1}{5}:\frac{3}{8}.$$

Kirk also saw that he could move multiplicatively from $^{1}/_{5}$ to $^{5}/_{5}$ by multiplying by 5, and he wrote that he could therefore multiply $^{3}/_{8}$ by 5 to obtain the solution of $^{15}/_{8}$ feet as the length of the original board. Clearly, he understood the multiplicative idea underlying the equivalence

$$\frac{1}{5}:\frac{3}{8}=\frac{5}{5}:\frac{15}{8}.$$

This idea is related to the ideas explored in Chapter 4. When first dividing by 2,

Kirk was setting up what might be characterized as a unit rate, in that he found that $1/5$ of the length of the board is $3/8$ foot long. This work is significant because it sets up a process that is easy to use with some algebraic equations. Algebraically, this situation is again expressed as

$$\frac{2}{5}x = \frac{3}{4}y,$$

and the focus is on solving for x. Kirk's method of determining the solution included two steps—first, dividing by the numerator 2, and second, multiplying by the denominator 5. This process is equivalent to multiplying both sides by the reciprocal of $2/5$, a technique used both in solving equations and in dividing fractions. That is, to solve the equation

$$\frac{2}{5}x = \frac{3}{4}y$$

for x, students might say, "Divide both sides by $2/5$." Kirk did this in a two-step process reflecting natural thinking within the context of the problem, and in doing so he demonstrated that dividing by $2/5$ produces a result equivalent to multiplying by $5/2$.

Hackenberg and Tillema (2005) make the case for fraction composition schemes as crucial when students begin to reason algebraically and as an important component in constructing meaning from and finding solutions to basic linear equations of the form $ax = b$. The reasoning demonstrated by Rochelle in the work shown in figure 5.8 extended the line of thinking used by Kirk to $ax = b$, where a and b are fractions. Thus, Rochelle moved further toward establishing a method for finding the solution $x = b/a$, which involves division of fractions, through a process similar to the commonly taught "invert and multiply" technique.

Rochelle saw that two and one-half $2/5$'s would make the whole board. Hence, she recognized that she needed to multiply $3/4$ by $2\,1/2$ to find the length of the board, which she conveniently wrote as

$$\frac{5}{2} \cdot \frac{3}{4} = \frac{15}{8}.$$

Rochelle demonstrated proportional thinking in constructing this equation, and her awareness and use of the relationship between $2\,1/2$ and $2/5$ should not be overlooked. Her work highlights the use of the reciprocal thinking discussed in Chapter 3 and earlier in this chapter. Furthermore, if the fraction $2/5$ in the problem had instead been $3/7$, so that $3/7$ of the board was $3/4$ foot long, Rochelle could have seen that two and one-third $3/7$'s, or $7/3$, would be needed to make the length of the whole board. When given $3/_7$, one might envision $2^1/_3$ in various ways, but in any case, these two values are reciprocals of each other.

This thinking, like Kirk's, leads to a generalization for solving for x in the equation

$$\frac{2}{5}x = \frac{3}{4}y$$

by multiplying both sides the equation by $5/_2$. Similarly, as discussed earlier, dividing both sides of the equation by $2/_5$ solves the equation for x,

$$x = \left(\frac{3}{4} \div \frac{2}{5}\right)y,$$

providing a rationale for the assertion that dividing by a fraction is the same as multiplying by the reciprocal of that fraction.

Canada, Gilbert, and Adolphson (2008) discuss thinking related to the proportional situation presented below as problem 5.3. They describe different ways of reasoning about and making sense of the situation that can be generalized into processes for solving proportion problems and can support multiplicative and building-up methods. Reflect 5.3 provides an opportunity to consider these suggestions, presented as ideas from three students for approaching and evaluating the proportional relationship embedded in the situation.

Problem 5.3

Alison and Stephanie were sent to buy ice cream for a party. Their favorite flavors came in a 64-ounce container for $6.79 and a 48-ounce container for $4.69. When they brought this information back to the people at the party, several people offered suggestions for ways to determine which was the better buy. (Adapted from Canada, Gilbert, and Adolphson [2008])

Reflect 5.3

Figures 5.9, 5.10, and 5.11 show, respectively, suggestions offered by three students, Alison, Sheryl, and Keith, of ways to work with the proportional relationship in problem 5.3 to determine which of the two containers of ice cream described is the better deal.

Explain how the suggestions by the three students would or would not help in determining which ice cream is the better buy.

How might the suggestions help to provide general methods for solving proportion problems?

$$\frac{64}{\$6.79} = 9.4256$$

$$\frac{48}{\$4.69} = 10.2345$$

Fig. 5.9. Alison's suggestion for approaching problem 5.3

$$\frac{64}{\$6.79} = \frac{64+32}{6.79+3.395}$$

$$\frac{48}{\$4.69} = \frac{48+48}{4.69+4.69}$$

Fig. 5.10. Sheryl's suggestion for approaching problem 5.3

$$\$6.79 \times 48 = \$325.92$$

$$\$4.69 \times 64 = \$300.16$$

Fig. 5.11. Keith's suggestion for approaching problem 5.3

Students often have difficulty with the approach that Alison suggests because they do not know how to interpret or compare 9.4256 and 10.2345. These computations are related to rates and represent ounces per dollar for the ice cream in each of the two containers, just as the more familiar rates, dollars per ounce, are given for each container by

$$\frac{\$6.79}{64} \text{ and } \frac{\$4.69}{48},$$

which yield \$0.1061 and \$0.0977 per ounce, respectively, for the 64-ounce and the 48-ounce container. By comparing cost per ounce, students can determine that the 48-ounce container of ice cream is cheaper per ounce, and therefore a better buy than the 64-ounce container of ice cream.

But they can make the same determination by comparing 9.4256 and 10.2345 if they think of these numbers as a result of changing each initial ratio into an equivalent ratio with a denominator of 1. In the case of the 64-ounce container of ice cream, for example,

$$\frac{64}{\$6.79} = \frac{9.4256}{1},$$

although writing 9.4256 ounces for $1 as

$$\frac{9.4256}{1}$$

is unconventional. Yet a ratio of 10.2345 ounces of ice cream to 1 dollar of cost clearly makes the 48-ounce container a better deal than the 64-ounce container, with its ratio of 9.4256 ounces of ice cream to 1 dollar of cost.

Thus, Alison's suggestion provides information that indicates that $4.69 for 48 ounces is the better buy because the number of ounces purchased for $1 is greater. Notice—and help your students notice, as well—the connection to reciprocal reasoning: you can look at given ratios or their reciprocals to answer a given question.

Sheryl suggests using a building-up strategy to produce two equivalent ratios that can be compared easily and directly because of the resulting common numerators. Making equivalent ratios of ounces of ice cream to cost, Sheryl determined 64 : 6.79 = 32 : 3.395 for the ice cream sold in the 64-ounce container, and then she used a building-up strategy to arrive at 64 : $6.79 = (64 + 32):(6.79 + 3.395), or 96 : 10.185. Similarly, she applied a building-up strategy to the ice cream sold in the 48-ounce container to show 48 : $4.69 = (48+48):(4.69+4.69), or 96 : 9.38. In other words, Sheryl saw how to build up the ratio relationships to compare 96 ounces of each of the two ice creams. By comparing 96 : 10.185

$$\left(\frac{96}{10.185}\right)$$

and 96 : 9.38

$$\left(\frac{96}{9.38}\right),$$

she can determine that the ice cream sold in the 48-ounce size is the better buy.

Sheryl could have also used a building-up strategy to find equivalent ratios with common denominators. This approach to the situation in problem 5.3 is similar to other ideas explored in Chapters 2 and 3, and earlier in Chapter 5, except that Sheryl chose to make the number of ounces in the new ratios constant—96 ounces in this case. This strategy emphasizes reasoning that can occur when common numerators are used to compare and analyze problems (Olson and Olson 2013; Olson,

Zenigami, and Slovin 2008). Using this method, students might see that dividing both quantities in the ratio

$$\frac{64}{\$6.79}$$

by 4, and dividing both quantities in the ratio

$$\frac{48}{\$4.69}$$

by 3, would produce new ratios, both of which in fraction form have a numerator of 16.

Keith's suggestion is similar to Atsuko's work shown in figure 5.3. Recall that Atsuko compared $^{14}/_{29}$ and $^{15}/_{31}$ by writing equivalent fractions with a common denominator of 29 × 31. Also recall that she saw this product as unnecessary to compute, focusing instead on 14 × 31 (434) and 15 × 29 (435)—the "cross products"—to identify $^{15}/_{31}$ as the larger fraction. Like Atsuko's work, Keith's suggestion points to the usefulness of cross multiplication to determine which ice cream is the better buy. He suggests comparing the ratios

$$\frac{\$6.79}{64} \text{ and } \frac{\$4.69}{48}$$

by looking at the products 6.79 × 48 (325.92) and 4.69 × 64 (300.16). Even though students can immediately see that 325.92 is greater than 300.16, they still must interpret these numbers to determine which ice cream is the better buy. A look back at the way that Atusko developed her information, as shown in the sample of her work in figure 5.3, suggests a general way of reasoning that can be applied to determine that 325.92 and 300.16 represent price per (64 × 48) ounces of the ice cream sold in the 64-ounce container and the ice cream sold in the 48-ounce container, respectively. From this information, students can readily determine the solution.

These three examples provide opportunities to explore techniques for working with ratio and proportion in different ways, all of which lead to processes that can be generalized. They illustrate the use of the building-up strategy to make equivalent ratios and also show how the means-extremes product property is useful in solving ratio problems.

Summarizing Pedagogical Content Knowledge to Support Essential Understanding 9

Teaching the mathematical ideas in this chapter requires specialized knowledge related to the four components presented in the Introduction: learners, curriculum, instructional strategies, and assessment. The four sections that follow summarize some examples of these specialized knowledge bases in relation to Essential Understanding 9. Although we separate them to highlight their importance, we also recognize that they are connected and support one another.

Knowledge of learners

The effectiveness of the thinking represented in Alison's, Sheryl's, and Keith's suggestions, shown in figures 5.9, 5.10, and 5.11—especially the latter two—resonates with suggestions about the importance of effective models offered by Fishbein and colleagues (1985). They found that unless students use effective models in their early explorations of problems involving particular concepts, they will continue to work with and adopt methods that will come into conflict with the formal, abstract reasoning expected in higher mathematics. In this case, unless students use effective models in solving problems involving ratios and proportional reasoning, they may flounder in solving linear equations. Alison's, Sheryl's, and Keith's suggestions demonstrate that students do understand models that connect well with the abstract models that come later, so building on them in instruction makes sense.

Often, focusing on the representations that students use is considered to be of less value than moving them to more abstract ways of demonstrating their understanding. However, Wu (2011) offers a valid and important observation as he discusses a model used by a student to solve a problem involving fractions: "We see plainly that there is no need to use multiplication of fractions for the solution, and moreover, no need to memorize any solution template. The present method of solution [a diagram followed by an explanation of what the diagram represented] makes the reasoning very clear" (pp. 36–37).

Knowledge of curriculum

Reporting on his own work and that of others, Smith (2002) indicates that even after students are taught cross multiplication, they rarely use it to solve missing-value problems. He elaborates (p. 16):

> Although cross-multiplication is both efficient and universally applicable (i.e., it does not depend on specific numbers in the problem), students either do not easily learn it or resist using it when they do. This may be due to the difficulty of linking this method to their early knowledge of ratios. The procedure does not match the mental operations involved in the "building up" strategy, and more specifically, the cross-products (e.g., "4 china × 35 silverware") lack meaning in the situation. What, after all, is one china-silverware, much less 140 of them?

Yet, as the examples in Chapter 5 demonstrate, students who can reason can solve ratio and proportion problems in various ways by making sense of the situation. At the same time, the choice of task can have an impact on the solution methods that your students use.

Knowledge of instructional strategies

Some believe that students have the most difficulty comparing fractions and ratios that lack multiples within or between themselves. Lamon (2001) shows work from two of her students on the question, "Are $^3/_5$ and $^7/_{11}$ equivalent?" Then she remarks, "Our results suggest that the kind of instruction children receive may have more impact on their ability to compare these fractions than the kind of numbers involved" (p 162). The type of instruction that serves students best builds on the reasoning of the students themselves and leads toward generalizations that become useful in the abstract at some point—although care must be taken before students have reached that point.

Several examples in Chapter 5 illustrate how students' reasoning related to ratio and proportion can build toward rules and procedures—but always from a sense-making perspective. In particular, Atsuko's work on problem 5.1, shown in figure 5.3, and Keith's suggestion for problem 5.3, shown in figure 5.11, provide a perspective on the means-extremes product property in relation to the cross-multiplication process. Sheryl's suggestion for problem 5.3, shown in figure 5.10, offers an opening for considering the power of comparing ratios with common numerators. Like Sheryl's thinking, the reasoning involved in Scott's use of reciprocals to solve problem 5.1, shown in figure 5.4, is not often found in textbooks but can support powerful strategies that students can employ to solve ratio and proportion problems. The samples of students' explanations for their work on problem 5.2 provide a way of explaining and justifying division of fractions as multiplying by reciprocals and using this strategy to solve simple linear equations when the coefficient of the variable is a fraction. Further, this strategy originates in a setting in which sense making leads directly to its development.

Knowledge of assessment

The examples from Canada, Gilbert, and Adolphson (2008) provide suggestions for formative assessment during instruction. Rather than focusing only on the solution, these suggestions draw students' attention to the meanings associated with various actions. Examining the reasoning behind the values obtained can lead to excellent discussions of ideas related to proportional thinking. These examples allow students to explore claims about the structures of relationships and how quantities do or do not vary as they are changed, while also allowing the students to use multiplicative relationships between two quantities.

The student work in Chapter 5 suggests that there are several important ways to think about proportion problems, including making connections with fractions and comparisons of fractions. The samples of student work also demonstrate that students can more readily understand the usual methods of solving word problems by applying concepts for finding equivalent ratios in a proportional context than by having these methods spelled out for them. Students should be given multiple opportunities to engage in problems of these types and to explain their generalizability.

Conclusion

To support you as you help your students develop several ways of reasoning, all grounded in sense making, that they can generalize into algorithms for solving proportion problems, this chapter has provided examples of student work and reflection questions that highlight ideas for putting Essential Understanding 9 into practice.

Emma's, Kirk's, and Rochelle's robust thinking, shown in figures 5.6, 5.7, and 5.8, respectively, about the length of Jonnine's board in problem 5.2, demonstrate that students can work with proportional settings without the need to use formal tools of proportionality to solve the problems. Furthermore, students can generalize the representational methods used by Kirk and Rochelle and use them in solving proportional problems in an algebraic setting. Working with and elaborating on these examples will prove fruitful when students confront other ratio and proportional settings, including scaling and covariance, in later studies.

into practice

Chapter 6 Looking Back and Looking Ahead with Ratios and Proportions

This chapter highlights ways in which the essential understandings discussed in Chapters 1–5 align with ideas related to ratios and proportions that students develop before and after grades 6–8. The first part of the chapter focuses on gaps that may be present in students' thinking about key ideas and examines several categories of ideas that students can build on in grades 3–5. The second part of the chapter highlights how the essential understandings related to ratios and proportions connect with, and can serve as an anchor for, many topics in mathematics that students will learn in mathematics beyond grade 8. Although these topics are too numerous to discuss exhaustively, the discussion illustrates several that serve as a foundation for subsequent learning in high school mathematics.

Building Foundations for Ratios and Proportions in Grades 3–5

Students' ability to reason about ratios and proportions draws heavily on their experiences with number and fraction before grade 6. Seven categories of related ideas help students to work effectively with ratios and proportions in grades 6–8, as discussed below.

1. The concept of unit, fractions as measurements, and "shifting units." Although students progress to thinking of a fraction as a number, initially they think of fractions in various ways. Whether they are using a linear, area, or set model for a fraction, students need to be aware of the unit under consideration. Students often think of fractions, especially when examining length or area, as measurements, such as $^3/_4$ of the length of a line segment or $^2/_5$ of the area of a square. As they progress to work with ratios, they extend the idea of measurement. For example, in using 6:18 ($^6/_{18}$) to represent the ratio of 6 cans of orange juice to 18 cans of water in a mixture, they learn that they are representing a measurement of "orange juiceness." This mixture may be too strong, too weak, or just right, depending on their desired taste. Similarly, a mixture of nuts that contains 5 cans of peanuts to 3 cans of almonds implies a measurement, whether students consider the measurement 5:3, 5:8, or 3:8. They come to understand that each of these ratios has an interpretation that represents the same relationship, but each is from a particular point of reference—peanuts to almonds, peanuts to total mixture, or almonds to total mixture—and each has an associated fraction that describes something about the relationship.

Attending to the frame of reference for the unit under consideration is also an important elementary mathematics experience. When students use the operations of multiplication and division, they find that units often "shift." For example, they can see a "shift" in a whole number division situation. Consider the following problem:

> Suppose that you have 42 pieces of candy and plan to give 7 pieces to each of several friends. To how many friends can you give candy?

In acting out this problem, students might use a single piece of candy as the unit as they move pieces aside to make a group of 7, but they might then consider the group of 7 pieces as a unit as they determine the solution to the problem. Similarly, if students are asked to find how many recipes can be made from 8 cups of flour when a single recipe calls for $^3/_4$ cup of flour, they will see that the answer to the question "What is the unit?" shifts from a unit of one cup to a unit of $^3/_4$ cup as they consider aspects of the problem.

2. Multiple models and representations. Students should learn how to use area or linear models, which are continuous, as well as discrete models, for fractions. CCSSM treats a linear representation as a preferred model for fractions. Linear models are also much easier for students to create than are the commonly used circle models. Even when students are using rectangular paper strips or drawings, they often rely on the attribute of linearity—not area. Although ratios and proportions may be represented by either discrete or continuous models, the use of continuous quantities is of special importance for most problems involving rates.

3. Comparison of quantities within and between fractions. When asked to decide which of two fractions is larger, students should be able to reason about the relationships of the numerators or the denominators, or use some other logical process. Students should be able to reason that if $a < b$, then for a number $c > 0$,

$$\frac{a}{c} < \frac{b}{c}.$$

They must also learn to reason that if $a < b$, then for a number $c > 0$,

$$\frac{c}{a} > \frac{c}{b}.$$

That is, students should be able to determine whether $^3/_5$ or $^3/_7$ is larger as well as whether $^3/_5$ or $^4/_5$ is larger. Being able to compare fractions when the numerators are the same is as important as being able to compare them when the denominators are the same. Likewise, when reasoning about ratios, students should be able to compare and contrast relationships between ratios such as 3:5 and 3:7 as well as ratios such as 3:5 or 4:5. Recall that for the mixture of nuts consisting of 5 cans of peanuts to 3 cans of almonds, students might use the measurement 5:3, 5:8, or 3:8. Similarly, when examining a diagram like that in figure 6.1, often used to represent $^3/_5$, they should be able to see $^2/_5$ as well as $^3/_5$. In addition, in working with the same diagram, students should be able to see

$$\left(1\frac{2}{3}\right)\left(\frac{3}{5}\right) = 1 \quad \text{and} \quad \left(2\frac{1}{2}\right)\left(\frac{2}{5}\right) = 1.$$

In other words, looking carefully at the diagram in figure 6.1 provides them with an opportunity to see

$$\frac{5}{3} \times \frac{3}{5} = 1 \quad \text{and} \quad \frac{5}{2} \times \frac{2}{5} = 1.$$

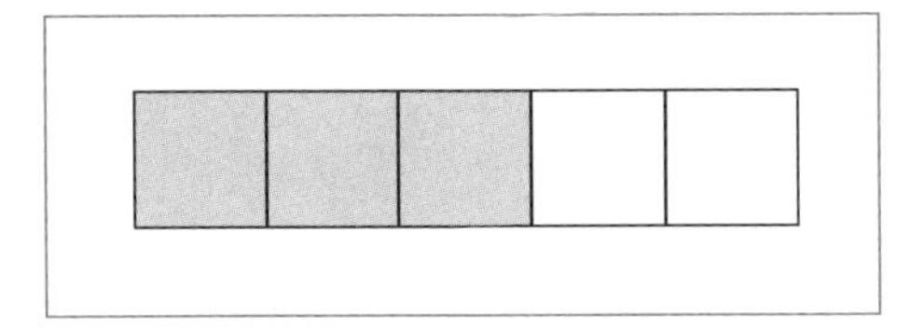

Fig. 6.1. Diagram of $^3/_5$

To think more closely about how students compare quantities within and between fractions and think in terms of units, consider problem 6.1. Use the questions in Reflect 6.1 to predict students' possible responses and reasoning, and then examine the work of seventh-grade students on the problem, shown in figures 6.2–6.6.

Problem 6.1

Which is closer to 1, $^5/_6$ or $^6/_5$?

Reflect 6.1

Can you think of several ways in which students might respond to problem 6.1?

What misconceptions would you expect to emerge in students' work?

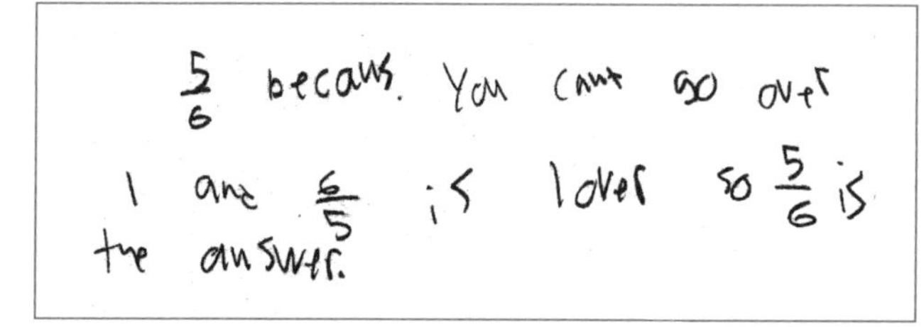

Fig. 6.2. Rico's response to problem 6.1

The fraction that is closer to is bouth
of them because there both the
same fraction but one is fliped
upside down.

Fig. 6.3. Ariel's response to problem 6.1

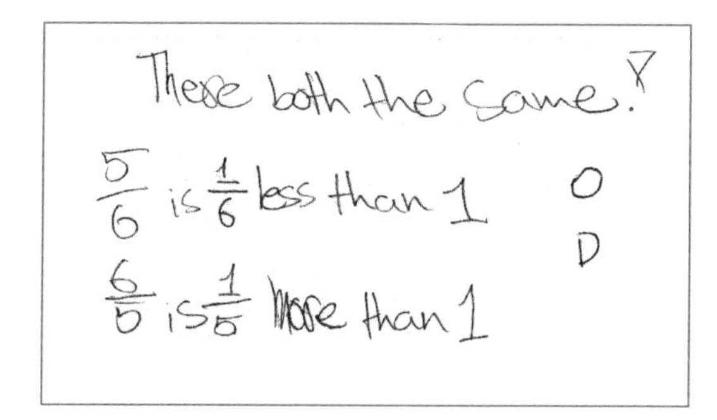

Fig. 6.4. Anne's response to problem 6.1

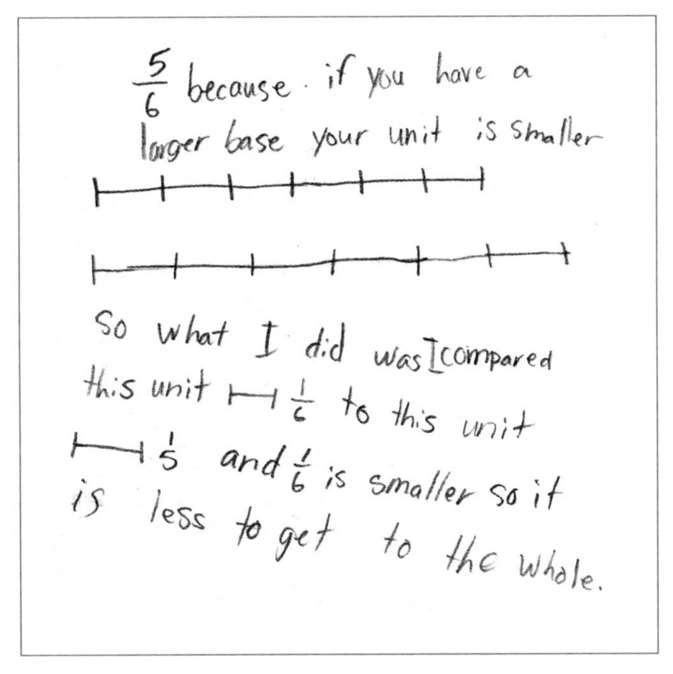

Fig. 6.5. Nikki's response to problem 6.1

5/6 is closer to 1 because 6/5 = 1 1/5
making it 1/5 away from one while
5/6 is 1/6 from one and compared
together is: 5/30 (1/6) and 6/30 (1/5), showing
that 1/6 is closer than 1/5 away.

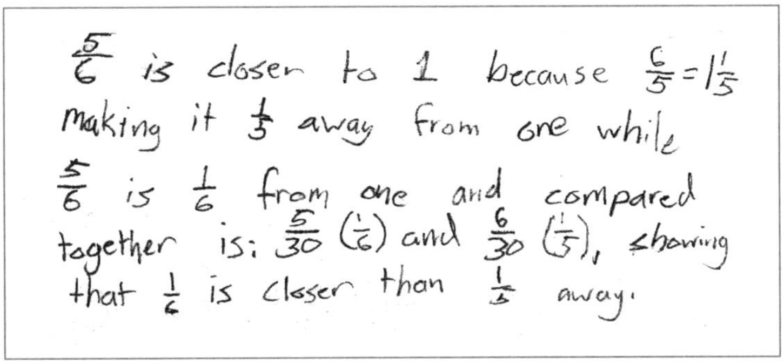

Fig. 6.6. Kevala's response to problem 6.1

Rico and Ariel certainly demonstrated some interesting perspectives on the question. Rico, whose work appears in figure 6.2, recognized that $^6/_5$ is "1 over," but he did not compare the relationship of $^6/_5$ to 1 in any meaningful fashion with the relationship of $^5/_6$ to 1, apparently because he thought that going "over" means that $^6/_5$ is not closer. It seems likely that Rico had previously worked with a situation in which being "over" when trying to get close to a number or value was not permissible.

Ariel's response, shown in figure 6.3, might reflect something previously learned as well. In thinking that the fractions $^5/_6$ and $^6/_5$ are the same but "one is flipped upside down," she might have been recalling features of the commutative properties for addition and multiplication, operations in which order does not matter.

By contrast, Anne's work, shown in figure 6.4, indicates that Anne recognized that comparing $1/6$ to $1/5$ is important, although she did not focus on the units involved but only on the 1's in the numerators. Consequently, she declared that the fractions are the same.

Nikki, whose work is shown in figure 6.5, had worked with units extensively, and although she did not document her thinking clearly, the two drawings that she made seem to have been helpful to her. The top drawing shows the unit broken into six pieces, whereas the bottom drawing has six segments, each of length $1/5$. She appears to have lined up the unit length quite well, assuming that the full segment on top represents one unit, and the fifth mark on the bottom segment represents the unit in the drawing below. Then she reasoned about the lengths representing $1/6$ and $1/5$ and concluded correctly that $6/5$ is closer to 1 than $5/6$.

Kevala, whose work appears in figure 6.6, also reasoned about $1/5$ and $1/6$, the distance between $6/5$ and 1 and $5/6$ and 1, respectively, on a number line. She justified the claim that $1/6$ is less than $1/5$ by renaming these fractions with common denominators and comparing $5/30$ and $6/30$, "showing that $1/6$ is closer than $1/5$ away."

These fractions were given to numerous middle-grades students for comparison, and those who used units and number lines were the most successful. When students think of $5/6$ as one-sixth less than 1 and $6/5$ as one-fifth greater than 1 and recognize that $5/6$ has to be closer to 1 because $1/6 < 1/5$, they compare fractions with greater proficiency.

4. Multiplicative thinking. The importance of multiplicative thinking for understanding fractions in grades 3–5 is discussed in *Putting Essential Understanding of Fractions into Practice in Grades 3–5* (Cheval, Lannin, and Jones 2013, pp. 110–14). The need to think multiplicatively continues and grows in grades 6–8, and middle-grades teachers must be careful not to assume that since students have studied multiplication, they think multiplicatively (Lamon 1999). To foster multiplicative thinking, students should be asked to look across as well as down entries in tables to determine patterns and complete entries.

Consider the table in figure 6.7, for example. If you ask your students to determine a relationship between the number of pencils and cost, they might look down the columns and note the numbers in the "Pencils" column increasing by 1 and the numbers in the "Cost in ¢" column increasing by 12. This process is useful not only for expressing recursive relationships but also for helping students learn to think in terms of covariation: as the values in the "Pencils" column increase by 1, the corresponding values in the "Cost in ¢" column increase by 12. However, your students also need to be able to look across the entries in the rows to determine that the values in a cost cell is 12 times the value in the corresponding pencil cell. Each of these ideas builds toward multiplicative thinking and also is related to the concept of slope.

Pencils	Cost in ¢
1	12
2	24
3	36

Fig. 6.7. Relationship of number of pencils and cost

5. Equivalence and nonequivalence. Students gradually learn that a fraction such as $^{2}/_{3}$ is one representation of an infinite collection of representations for the same value. They must also learn how to find a fraction that is equivalent to another fraction within given parameters and be able to justify their claim of equivalence. Problems 6.2 and 6.3 are examples of problems that can help students explore the nuances of equivalent fractions. Examine problem 6.2 first, and then consider the work of a fifth-grade student, Justin, presented in figure 6.8.

Problem 6.2

Find several fractions that when simplified are equivalent to $^{5}/_{6}$.

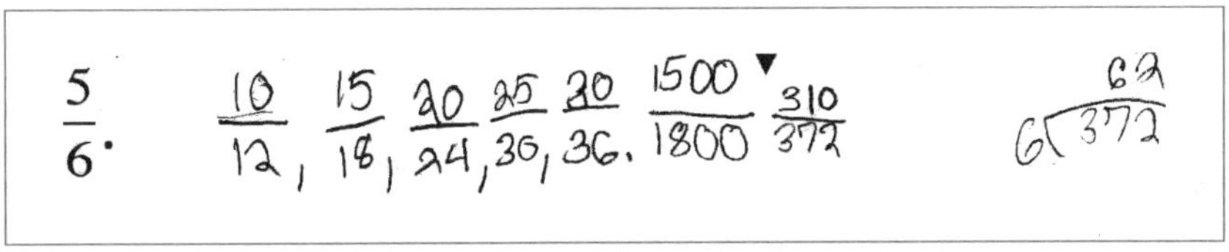

$\frac{5}{6}$. $\frac{10}{12}$, $\frac{15}{18}$, $\frac{20}{24}$, $\frac{25}{30}$, $\frac{30}{36}$. $\frac{1500}{1800}$ ▼ $\frac{310}{372}$ 62 6)372

Fig. 6.8. Justin's response to problem 6.2

Although Justin showed little of his reasoning, he was able to generate several such fractions. Justin's teacher might have elicited interesting information by asking Justin why after creating the first five examples, he offered $^{1500}/_{1800}$, and what he was thinking when he obtained $^{310}/_{372}$. Also consider problem 6.3 and the response from another fifth-grade student, Lisa, shown in figure 6.9.

Problem 6.3

Find all values of x, $x < 19$, such that $^{x}/_{30}$ can be simplified.

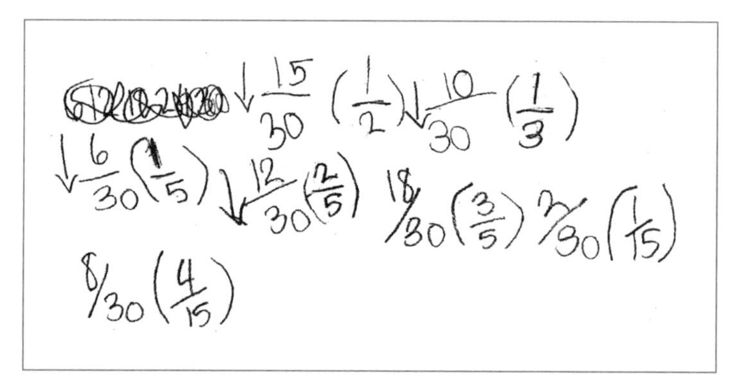

Fig. 6.9. Lisa's response to problem 6.3

In the short time allowed, Lisa identified seven of the thirteen possible values, along with the simplified form of each fraction. Problems like this one and the previous one allow students to explore equivalent fractions in a different manner while building other number skills.

Students must not rush to overgeneralize ideas. For example, they must not claim that $^{2}/_{3}$ is equivalent to $^{3}/_{4}$ because they are both one part less than a whole—an idea that is similar to the thinking in Anne's response, given in figure 6.4. Further, figure 6.10 shows comparative thinking, overgeneralized and used incorrectly by a seventh-grade student, Kelvin, in explaining whether the two fractions $^{6}/_{9}$ and $^{10}/_{15}$ are equivalent (work on this task by other students appears in Chapter 3).

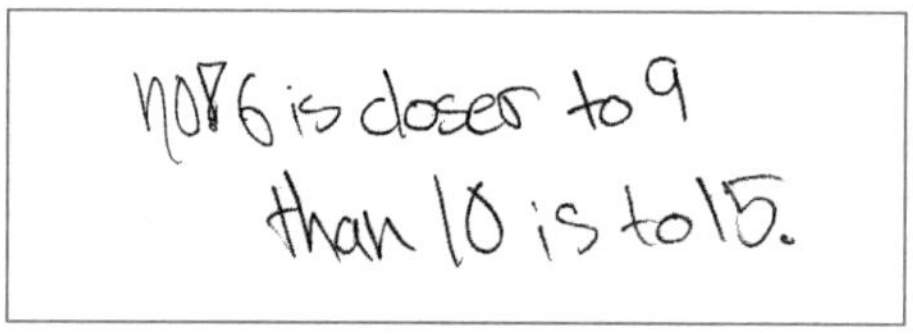

no, 6 is closer to 9
than 10 is to 15.

Fig. 6.10. Kelvin's evaluation of whether $^{6}/_{9}$ and $^{10}/_{15}$ are equivalent

6. Contextualization and decontextualization. "Reason abstractly and quantitatively" is standard 2 of the Standards for Mathematical Practice in the Common Core State Standards for Mathematics (CCSSM; National Governors Association Center for Best Practices and Council of Chief State School Officers [NGA Center and CCSSO] 2010). As noted in Chapters 2 and 5, this core mathematics practice incorporates essential paired abilities:

> Mathematically proficient students ... bring two complementary abilities to bear on problems involving quantitative relationships: the ability to *decontextualize*—to abstract a given situation and represent it symbolically and manipulate the representing symbols as if they have a life of their own, without necessarily attending to their referents—and the ability to *contextualize*, to pause as needed during the manipulation process in order to probe into the referents for the symbols involved. (NGA Center and CCSSO 2010, p. 6)

Students must become able to move from concrete representations to representations of fractions on a number line and to think of a fraction as a number. They may begin by thinking in terms of the units given in a context, but just as important, they must be able to put units to numbers if necessary. Students must be given problems in context as well as mathematical exercises, and when given a problem in context, they need to figure out how the numerical values emerge and are related. When given numbers without a context, they should try to envision a context with which these numbers might be associated, as illustrated by Adele's work, shown in Chapter 5 (fig. 5.5).

7. Flexible thinking. In working on problems, students who can reason well about equivalent fractions and about numbers in context have an advantage when they move to making sense of problems involving ratios. For example, reconsider the reasoning explained by Elise on problem 3.2, about the board from which Jonnine cut $^2/_5$, with the cut piece measuring $^3/_4$ of a foot. Elise's work to determine the length of the whole board appeared earlier as figure 3.11 and is shown again for convenience on the next page as figure 6.11.

Although the wording "$^2/_5$ equals $^3/_4$" is questionable, Elise gave a sound verbal explanation: if two-fifths of the board measured $^3/_4$ of a foot, then four-fifths would be $^6/_4$ of a foot. Also, if two-fifths of the board measured $^3/_4$ foot, then one-fifth would be half of that, which Elise deduced would be $1^1/_2$ fourths of a foot. So the whole board would measure $^6/_4$ of a foot plus $1^1/_2$ fourths of a foot. That is, she understood the structure of the problem and could think flexibly about how to find one-half of $^3/_4$, and she was sufficiently confident and comfortable in her thinking to write her solution in a form that made sense, even if she did not write it in the conventional way.

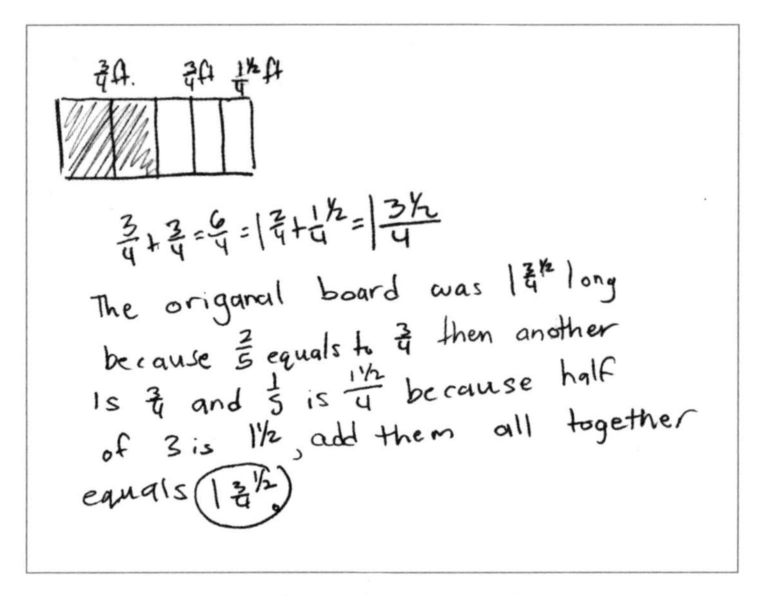

Fig. 6.11. Elise's solution to problem 3.2

Elise's thinking is similar to reasoning that Olson and Zenigami (2009) report from a student, Michael, when he was asked to simplify

$$\frac{35}{40} \quad \text{and obtained} \quad \frac{7\frac{3}{4}}{10}.$$

He explained that he divided 35 and 40 by 2, and then took the results and divided each of them by 2. He tried to do it again but couldn't, so he stayed with

$$\frac{7\frac{3}{4}}{10}$$

as the simplified form. Although this response was not the expected one (and Michael made an error in his division of 35 by 2 and 2 again), the process demonstrated flexible thinking on Michael's part.

Extending Understanding of Ratios and Proportions in Grades 9–12

Writing more than 70 years ago, Cowley (1937) discussed the significance of the concepts of ratio, proportion, and variation. He examined high school algebra, geometry, chemistry, and physics books "to study the high school pupils' grasp of relationship of measurable quantities, as revealed by the comprehension of ratio and proportion" (p. 1079). CCSSM addresses five specific content areas in grades 9–12—quantities, equations, functions, geometry, and statistics—in which the use of ratios, proportions, and proportional reasoning extends the ideas explored in grades 6–8. These areas do not encompass every specific bit of content to which the ideas apply. The discussion that follows focuses on two key areas, invariance and transformations, in which concepts of ratio, proportion, and proportional reasoning are connected and foundational. Although these two areas are not disjoint, the discussion that follows presents separate examples that highlight relevant content in each.

1. Invariance

Students apply their understanding of invariant relationships in ratios and proportions to many topics in high school mathematics. The examples below from the content areas of functions and algebra highlight just two applications.

Example 1. CCSSM high school conceptual category: Functions

Domain: Interpreting Functions

Cluster: Analyze functions using different representations

Standard:

8. Write a function defined by an expression in different but equivalent forms to reveal and explain different properties of the function. (F-IF.8; NGA Center and CCSSO 2010, p. 69)

Students examine functions that arise in many settings. Suppose that they want to find the number of toothpicks that they would need to make any number of squares configured as in figure 6.12. For example, for the 6 squares shown, students would use 19 toothpicks.

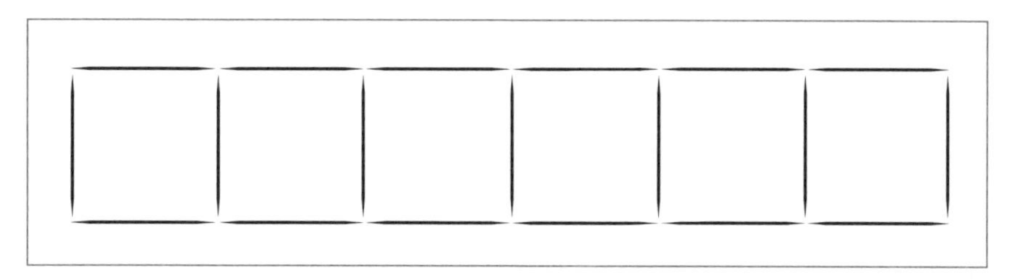

Fig. 6.12. Toothpick diagram

When students are asked to determine an expression for the number of toothpicks (y) for any number of squares (x), they often obtain the expression $y = 1 + 3x$, by seeing the first vertical toothpick with toothpicks added to it in groups of 3 for each new square. In this case, 3 is the value of the slope of the equation. Thus, it is an invariant added at each step, and although the line does not show a proportional relationship, 3 is the ratio representing the corresponding change in y-values for associated changes in x-values.

Example 2. CCSSM high school conceptual category: Algebra

Domain: Creating Equations

Cluster: Create equations that describe numbers or relationships

Standard:

4. Rearrange formulas to highlight a quantity of interest, using the same reasoning as in solving equations. *For example, rearrange Ohm's law $V = IR$ to highlight resistance R.* (A-REI.4; NGA Center and CCSSO 2010, p. 65)

In the equation given in CCSSM, when students isolate R as $R = {}^{V}/_{I}$, they see that for a given value of R (say 20), V and I are directly proportional because the ratio of V to I is always 20. If R is constant (invariant), then students understand that V and I are directly proportional in that as one increases by a specific factor, so does the other. Similarly, if students isolate I and hold it constant (invariant), they see that V and R are directly proportional.

This situation also provides teachers with an opportunity to discuss variation that is not always directly proportional. For example, if V is held constant (invariant), I and R are inversely proportional because as one increases by a factor, the other decreases by the same factor. Boyle's law relating gas pressure and volume of gas at a constant temperature is another example of inverse variation.

2. Transformations

High school mathematics students frequently apply proportional reasoning as they think about, and work with, transformations and the properties of transformations. The following two examples from the content areas of geometry and algebra illustrate aspects of this work.

Example 1. CCSSM high school conceptual category: Geometry

Domain: Similarity, Right Triangles, and Trigonometry

Cluster: Prove theorems involving similarity

Standard:

5. Use congruence and similarity criteria for triangles to solve problems and to prove relationships in geometric figures. (G-SRT.5; NGA Center and CCSSO 2010, p. 77)

A well-known theorem in geometry states that if a line is parallel to one side of a triangle and intersects the other two sides, then it divides those sides proportionally.

For example, in triangle ABC in figure 6.13, DE is parallel to BC, with D on AB and E on AC, and hence,

$$\frac{AD}{DB} = \frac{AE}{EC}.$$

(Note that although the discussion is about proportionality and thus involves ratios, the relationship is commonly written in "fraction form," as above. It could also be written as $AD{:}DB = AE{:}EC$).

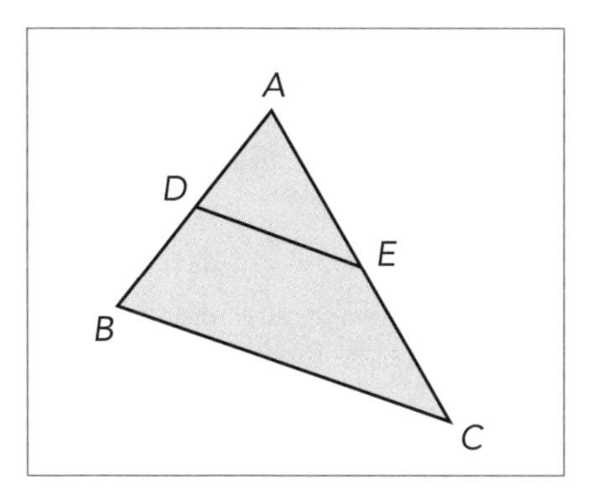

Fig. 6.13. Similar triangles

However, if students use dilation, they find that this relationship is true by properties of dilations. If k is the dilation that maps segment AD to segment AB and segment AE to segment AC, then segment DE is parallel to segment BC by properties of dilations. Also, because of the dilation constant k,

$$\frac{AD}{AB} = \frac{AE}{AC}.$$

From this proportion, students can also derive

$$\frac{AD}{DB} = \frac{AE}{EC}$$

and several other proportions associated with similar triangles.

Students can use similar reasoning associated with dilations to prove the corollary that states that when three parallel lines intersect two transversals, the segments intercepted on the transversals are proportional. For example, in figure 6.14,

$$\frac{BC}{CD} = \frac{XY}{YZ}.$$

Students can see that this ratio generates several other proportional relationships as well.

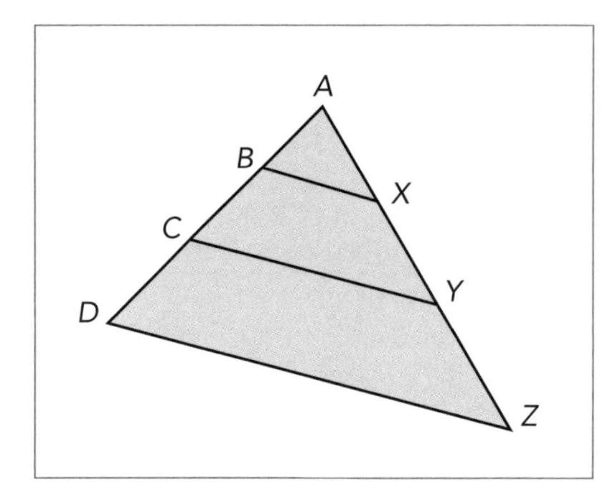

Fig. 6.14. Three parallel lines with two transversals

Example 2. CCSSM high school conceptual category: Algebra

Domain: Reasoning with Equations and Inequalities

Cluster: Represent and solve equations and inequalities graphically

Standard:

10. Understand that the graph of an equation in two variables is the set of all its solutions plotted in the coordinate plane, often forming a curve (which could be a line). (A-REI.10; NGA Center and CCSSO 2010, p. 66)

In addition to understanding that non-vertical lines that go through the origin express proportional relationships, students should understand that non-vertical lines that do not go through the origin do not express proportional relationships, but the slope of such a line nevertheless expresses a ratio that is meaningful. Beyond developing the understanding that CCSSM expects in standard 10, students should spend time exploring relationships between the graphs of two linear equations with the same slope. In working with the two equations $y_1 = {}^2/_3\, x + 7$ and $y_2 = {}^2/_3\, x$, for example, they should not only compare and contrast the slopes and y-intercepts but should understand how the y-values in the two equations differ for the same x-value. That is, they should know that the line $y_1 = {}^2/_3\, x + 7$ is a translation of $y_2 = {}^2/_3\, x$, and hence the lines are parallel by properties of translations. Although the number of translations of y_1 onto y_2 is infinite, one such translations maps (x, y) to $(x, y + 7)$.

Students can explore similar questions when equations are written in the general form. For example, they should compare and contrast the slopes and y-intercepts for equations such as $2x + 3y = 9$ and $2x + 3y = 15$ and also compare the ways in which the y-values in the equations differ for the same x-value. Students should also be able to identify at least one translation that maps one line onto the other. Knowing that translations map a line to another line that is parallel is a starting point for understanding that for two given lines, there is a transformation that maps one to the other.

Conclusion

Chapter 6 concludes with an opportunity for you to examine two situations that underscore the complexity of proportional reasoning and the difficulty that problems in context can create. Kahan and colleagues (2013) demonstrate just how complicated issues related to proportional reasoning can be. Even highly numerate individuals do not always interpret situations like the following correctly, as a result of beliefs that they hold about them. Use the questions in Reflect 6.2 and 6.3 to guide your thinking about situations 6.1 and 6.2, respectively, and then use the questions in Reflect 6.4 to help you analyze the difference between the two situations.

Situation 6.1

Suppose that a treatment is being tried to assist patients who have a particular disease. Data related to the treatment are organized below, with results compiled from both people who did and people who did not receive the treatment. Of those receiving the treatment, 115 became worse and 39 became better, and of those not receiving the treatment, 56 became worse and 11 became better:

Treatment	Worse	Better
Received	115	39
Did not receive	56	11

Reflect 6.2

Which result would the study in situation 6.1 support?

A. Those who were treated were more likely to get *better* than those who were not?

B. Those who were treated were more likely to get *worse* than those who were not?

Situation 6.2

Suppose that for the same treatment as in situation 6.1 the data are presented in the same manner as before, but with words "better" and "worse" exchanged. That is, suppose that of those receiving the treatment, 115 became better and 39 became worse, and of those not receiving the treatment, 56 became better and 11 became worse:

Treatment	Worse	Better
Received	115	39
Did not receive	56	11

Reflect 6.3

Which result would the study in situation 6.2 support?

A. Those who were treated were more likely to get *better* than those who were not?

B. Those who were treated were more likely to get *worse* than those who were not?

Reflect 6.4

How do the changes in the arrangement of the data matter? Compare and contrast your thinking related to situation 6.1 with your thinking related to situation 6.2.

Kahan and colleagues (2013) found that when the situation under consideration described a treatment with a relatively neutral emotional, cultural, or political value or association (for instance, moisturizing skin cream), highly numerate people answered each question at about the same level of competence. That is, by using whatever method they chose to compare ratio relationships such as 115 : 154 ($^{115}/_{154}$) and 56 : 67 ($^{56}/_{67}$), or odds such as 115 to 39 and 56 to 11, highly numerate people were able to obtain information from the data given in situations 6.1 and 6.2 and apply it accurately to select response A in the case of situation 6.1 and response B in the case of situation 2.

However, when highly numerate people had a very emotional, cultural, or political association with the treatment under consideration (for example, banning guns), those in favor of the treatment correctly selected response A in the case of situation 6.1, but failed to select response B in the case of situation 2. Similarly, highly numerate people opposing the treatment correctly selected response B in the case of situation 2, but failed to select response A in the case of situation 1. This illustrates that even when people demonstrate the proportional reasoning knowledge to solve a problem in context correctly, using that knowledge to choose an answer that does not align with their opinions may be hard for them. That is, they may not separate the context of the problem from their beliefs about what they are examining. Keep in mind when you are working with proportional reasoning with students that the setting in which you situate a problem may have an impact on a student's response.

Work with ratios, proportions, and proportional reasoning in grades 6–8 has connections with numerous topics in grades 9–12. Studying the invariance associated with direct variation is important in work with algebra, functions, geometry, and statistics. Similarly, using properties of dilations connects well with understanding relationships between lines and much of students' work with similar triangles.

Appendix 1

The Big Idea and Essential Understandings for Ratios, Proportions, and Proportional Reasoning

This book focuses on the big idea and essential understandings that are identified and discussed in *Developing Essential Understanding of Ratios, Proportions, and Proportional Reasoning for Teaching Mathematics in Grades 6–8* (Lobato and Ellis 2010). For the reader's convenience, the big idea and the complete list of essential understandings in that book are reproduced below. Those that are the special focus of this book are highlighted in blue.

Big Idea. When two quantities are related proportionally, the ratio of one quantity to the other is invariant as the numerical values of both quantities change by the same numerical factor.

Essential Understanding 1. Reasoning with ratios involves attending to and coordinating two quantities.

Essential Understanding 2. A ratio is a multiplicative comparison of two quantities, or it is a joining of two quantities in a composed unit.

Essential Understanding 3. Forming a ratio as a measure of a real-world attribute involves isolating that attribute from other attributes and understanding the effect of changing each quantity on the attribute of interest.

Essential Understanding 4. A number of mathematical connections link ratios and fractions:

- Ratios are often expressed in fraction notation, although ratios and fractions do not have identical meaning.
- Ratios are often used to make "part-part" comparisons, but fractions are not.

- Ratios and fractions can be thought of as overlapping sets.
- Ratios can often be meaningfully reinterpreted as fractions.

Essential Understanding 5. Ratios can be meaningfully reinterpreted as quotients.

Essential Understanding 6. A proportion is a relationship of equality between two ratios. In a proportion, the ratio of two quantities remains constant as the corresponding values of the quantities change.

Essential Understanding 7. Proportional reasoning is complex and involves understanding that–

- equivalent ratios can be created by iterating and/or partitioning a composed unit;
- if one quantity in a ratio is multiplied or divided by a particular factor, then the other quantity must be multiplied or divided by the same factor to maintain the proportional relationship; and
- the two types of ratios–composed units and multiplicative comparisons–are related.

Essential Understanding 8. A rate is a set of infinitely many equivalent ratios.

Essential Understanding 9. Several ways of reasoning, all grounded in sense making, can be generalized into algorithms for solving proportion problems.

Essential Understanding 10. Superficial cues present in the context of a problem do not provide sufficient evidence of proportional relationships between quantities.

Appendix 2
Resources for Teachers

The following list highlights a few of the many books, articles, and online materials that offer helpful resources for teaching ratios and proportions in grades 6–8.

Books

Lamon, Susan J. *Teaching Fractions and Ratios for Understanding: Essential Content Knowledge and Instructional Strategies for Teachers.* 3rd ed. New York: Routledge, 2012.

———. *More! Teaching Fractions and Ratios for Understanding: In-Depth Discussion and Reasoning Activities.* 3rd ed. New York: Routledge, 2012.

Lamon's book and its supplement highlight children's strategies for the reader to analyze. Both books provide activities for classroom use and offer reflection questions to help readers explore "between" and "within" strategies.

Litwiller, Bonnie, ed. *Making Sense of Fractions, Ratios and Proportions.* 2002 Yearbook of the National Council of Teachers of Mathematics. Reston, Va.: National Council of Teachers of Mathematics, 2002.

Bright, George, and Bonnie Litwiller, eds. *Classroom Activities for "Making Sense of Fractions, Ratios, and Proportions."* Reston, Va.: National Council of Teachers of Mathematics, 2002.

NCTM's 2002 Yearbook and its companion activity book combine to provide discussion for mathematics educators and hands-on activities and problems for students. Articles in the yearbook clarify issues related to fractions, ratios, and proportions, and activities in the companion book illustrate ways of engaging students with some of the ideas in the classroom.

Van de Walle, John A., Karen S. Karp, and Jennifer M. Bay-Williams. *Elementary and Middle School Mathematics: Teaching Developmentally.* 8th ed. Needham Heights, Mass.: Allyn & Bacon, 2013.

This resource is often used with preservice and in-service teachers to expand thinking related to a myriad of mathematics topics in K–8. The material on ratio, proportions, and rates is designed to develop teachers' thinking on these topics.

Articles

Canada, Dan, Mike Gilbert, and Keith Adolphson. "Investigating Mathematical Thinking and Discourse with Ratio Triplets." *Mathematics Teaching in the Middle School* 14 (August 2008): 12–17.

This article describes a unique task structure designed to foster classroom discourse and reveal students' mathematical conceptions. The example involves a proportional reasoning task set in a best-buy context dealing with purchasing ice cream for a party. The authors created three versions of the task and gave different versions to different groups of students. They then investigated the multiple ways in which the versions encouraged students to consider ratios and proportions. The article presents and analyzes several student responses.

Che, S. Megan. "Giant Pencils: Developing Proportional Reasoning." *Mathematics Teaching in the Middle School* 14 (March 2009): 404–8.

A problem about giant pencils encourages students to reason proportionally. A rich, open-ended mathematical task gives students opportunities to develop proportional reasoning and explore measurement. The article discusses various strategies that students might use to solve the problem and offers ideas for extending their thinking.

Cobbs, Georgia A., and Edith Cranor-Buck. "Getting into Gear." *Mathematics Teaching in the Middle School* 17 (October 2011): 160–65.

A classroom activity, the Motorized Toy unit, is at the center of this article. This team-taught activity supports STEM goals and teaches students the basic concept of ratio as they answer two questions in a driving context: "How far did your car climb?" and "What gear ratio did you use?"

Jarvis, Daniel. "Math Roots: Mathematics and Visual Arts: Exploring the Golden Ratio." *Mathematics Teaching in the Middle School* 12 (April 2007): 467–71.

The author presents the golden ratio in three contexts: historical, mathematical, and pedagogical. The article includes a brief history, various teaching strategies, and a project for middle school mathematics students.

Lamon, Susan J. "Rational Numbers and Proportional Reasoning: Toward a Theoretical Framework for Research." In *Second Handbook of Research on Mathematics Teaching and Learning*, edited by Frank K. Lester, Jr., pp. 629–68. Charlotte, N.C.: Information Age; Reston, Va.: National Council of Teachers of Mathematics, 2007.

This article takes the reader through past and current research on proportional reasoning, contains a wealth of examples with an analysis of student thinking and reasoning, and discusses, among other topics, covariance and invariance and multiplicative reasoning.

Lim, Kien. "Burning the Candle at Just One End." *Mathematics Teaching in the Middle School* 14 (April 2009): 492–500.

When students learn about proportions, they must understand what makes a situation proportional. This article illustrates the use of proportional and nonproportional situations to enable students to analyze a problem situation, determine the manner in which quantities covary, and identify the relationship that is invariant.

Mamolo, Ami, Margaret Sinclair, and Walter J. Whiteley. "Proportional Reasoning with a Pyramid." *Mathematics Teaching in the Middle School* 16 (May 2011): 544–49.

A three-dimensional model and geometry software help develop students' spatial reasoning and visualization skills and illustrate the importance of exposing students to a wide variety of contexts in which to represent proportional reasoning.

Reeder, Stacy. "Are We Golden? Investigations with the Golden Ratio." *Mathematics Teaching in the Middle School* 13 (October 2007): 150–54.

This article describes a project-based learning experience that involves students in investigating the golden ratio. Middle school students look at the "golden nature" of their bodies through measurement, data collection, and proportional reasoning. An activity sheet is included.

Roberge, Martin, and Linda Cooper. "Map Scale, Proportion, and Google Earth." *Mathematics Teaching in the Middle School* 15 (April 2010): 448–57.

Students work with aerial photographs to explore the concept of map scale and use proportional reasoning to analyze the relationship between the image size and the actual ground size of familiar objects.

Scaptura, Christopher. "Home Area and History." *Mathematics Teaching in the Middle School* 13 (February 2008): 349–50.

A question from a social studies context about how much living space a person needs leads students to conduct an investigation related to area, scale drawing, and ratio.

Slovin, Hannah. "Take Time for Action: Moving to Proportional Thinking." *Mathematics Teaching in the Middle School* 6 (September 2000): 58–60.

This article explores how students make sense of ratio and proportion through the use of dilation.

Snow, Joanne E., and Mary Kay Porter. "Math Roots: Ratios and Proportions: They Are Not All Greek to Me." *Mathematics Teaching in the Middle School* 14 (February 2009): 370–74.

This article examines the history of ratios and proportions and their value to people from the past through the present.

Online Resources

Dan Meyer's Three-Act Math Tasks
threeacts.mrmeyer.com

This website provides numerous opportunities to use the "three act" format for engaging students. The first act involves students in watching a video, often without words, from which they generate and formulate a question or questions. The second act engages them in investigating solution strategies, often with a strong modeling component, to arrive at a solution. The third act reveals the solution or range of potential solutions. Several activities allow students to explore ideas and strategies related to ratio and proportion, including the "within" or "between" strategies, as they approach the problem from the perspective with which they are most comfortable. Allowing students to select strategies and work from their own perspective in this way gives teachers an opportunity to see the range of ways in which students attend to invariance in the given situations.

Illuminations
http://illuminations.nctm.org/

The NCTM Illuminations project originated as part of the Verizon Thinkfinity program and continues to develop and present a variety of standards-based resources, including lessons, activities, and hundreds of Web links. The following Illuminations lessons specifically address ratio and proportion:

The Golden Ratio
http://illuminations.nctm.org/Lesson.aspx?id=1658

Students learn about ratios, including the "golden ratio," a ratio of length to width that can be found in art, architecture, and nature. Students examine different ratios to determine whether the golden ratio can be found in the human body.

Highway Robbery
http://illuminations.nctm.org/Lesson.aspx?id=3128

The National Bank of Illuminations has been robbed! Students apply their knowledge of ratios, unit rates, and proportions to sort through the clues and deduce which suspect is the culprit.

In Your Shadow
http://illuminations.nctm.org/Lesson.aspx?id=1672

Students extend their knowledge of proportions to solving problems dealing with similarity. They measure the heights and shadows of familiar objects and use indirect measurement to find the heights of things that are much bigger, such as a flagpole, a school building, or a tree.

Feeding Frenzy
http://illuminations.nctm.org/Lesson.aspx?id=2854

Students multiply and divide a recipe to feed groups of various sizes. They use unit rates or proportions and think critically about real-world applications of a baking problem.

Understanding Rational Numbers and Proportions
http://illuminations.nctm.org/Lesson.aspx?id=1110

Students use real-world models to develop an understanding of fractions, decimals, unit rates, proportions, and problem solving. This investigation features three activities that center on situations involving rational numbers and proportions at a bakery. The activities involve several important concepts of rational numbers and proportions, including partitioning a unit into equal parts, the quotient interpretation of fractions, the area model of fractions, determining fractional parts of a unit not cut into equal-sized pieces, equivalence, unit prices, and multiplication of fractions.

Bean Counting and Ratios
http://illuminations.nctm.org/Lesson.aspx?id=2534

By working with samples drawn from a large collection of beans, students get a sense of equivalent fractions, and this leads to a better understanding of proportions. Equivalent fractions are used to develop an understanding of proportions. Teachers can adapt the lesson for younger or less-skilled students by using a more common ratio, such as $^{2}/_{3}$, and for older or more skilled students by using a less familiar one, such as $^{12}/_{42}$ (equivalent to $^{2}/_{7}$).

Shops at the Mall
http://illuminations.nctm.org/Lesson.aspx?id=1049

Students develop number sense by engaging in an activity set in and around a shopping mall. They solve problems involving percentage and scale drawings.

Off the Scale
http://illuminations.nctm.org/Lesson.aspx?id=1675

Students use real-world examples to solve problems involving scale as they examine maps of their home states and calculate distances between cities.

The Ratio of Circumference to Diameter
http://illuminations.nctm.org/Lesson.aspx?id=1849

Students measure circular objects. They calculate the ratio of circumference to diameter for each object as they work to determine the value of pi and the formula for circumference.

Integrating Mathematics and Pedagogy (IMAP)
http://www.sci.sdsu.edu/CRMSE/IMAP/video.html

The IMAP website integrates information about children's mathematical thinking into mathematics content for preservice elementary teachers. The site features presentations, publications, and video clips of children doing mathematics. Additional video clips are available on *IMAP: Select Videos of Children's Reasoning*, a CD that contains twenty-five clips of elementary school children engaged in mathematical thinking. The CD runs on both PC and Mac platforms and comes with an interface that includes the transcript (full or synchronized) and background information for each clip. Also included on the CD is a video guide containing questions to consider before and after viewing each video clip, interviews that teachers or prospective teachers can use when working with children, and other resources.

LearnZillion
http://learnzillion.com/lessons

LearnZillion is a learning platform that combines video lessons, assessments, and progress reporting for mathematics and English language arts. Each lesson highlights a Common Core standard (currently grades 3–9 in mathematics). Classroom teachers working with coaches created the lessons and materials on this site. The online lessons consist of PowerPoint slides with narratives lasting approximately five minutes. Each lesson addresses a specific standard and focuses on the conceptual understanding that students are expected to have for that standard. Although the lessons are not intended to constitute a curriculum, they could serve as useful introductions or warm-up activities for lessons or as concluding exercises at the end of lessons.

The Math Forum
http://mathforum.org/mathtools/

The Math Forum is a rich resource for teachers, kindergarten–grade 12. The site features MathTools, a digital library of mathematics resources, including a catalog of materials and a discussion forum, to support mathematics instruction and learning. Topics for grades 6 and 7 include ratio and proportion, with numerous problems and activities.

Common Core State Standards for Mathematics (CCSSM) and Closely Related Resources

Common Core State Standards for Mathematics (CCSSM)
http://www.corestandards.org/Math

Starting with the standards themselves is a good way to strengthen and guide curricular decisions.

CCSSM Curriculum Materials Analysis Tools
http://www.mathedleadership.org/ccss/materials.html

The website of the National Council of Supervisors of Mathematics offers resources to help with the implementation of CCSSM, kindergarten–grade 12. In particular, it offers a tool that can help teachers determine the alignment of curricular materials with CCSSM.

Common Core Projects at IM&E
http://ime.math.arizona.edu/commoncore

The Institute for Mathematics and Education (IM&E) at the University of Arizona offers resources that amplify the meaning of the Common Core State Standards for Mathematics and align instruction with them. Bill McCallum, a mathematician at the University of Arizona and lead member of the CCSSM writing team, maintains this comprehensive website.

Illustrative Mathematics
http://www.illustrativemathematics.org/

Resources available at this site illustrate the range and types of mathematical work that students experience in a faithful implementation of the Common Core State Standards. The website also provides tools that support implementation of the standards—tasks and videos that show what *precision* means at various grade levels.

Learning Trajectories for the K–8 Common Core Math Standards
http://turnonccmath.net

This website unpacks CCSSM for kindergarten–grade 8 in eighteen learning trajectories, developed through the work of Jere Confrey. The learning trajectories show how the understanding that students develop in elementary and middle school serves as the foundation for high school study of functions. TurnOnCCMath.net presents these trajectories in a set of graphics and narratives that connect the K–8 standards across the grade bands. The Generating Increased Science and Math Opportunities (GISMO) research team at North Carolina State University identified gaps in CCSSM and developed bridging standards to span those gaps. Clicking on the hexagons in the map at the website gives access to the bridging standards.

Progressions Documents for the Common Core Math Standards
http://ime.math.arizona.edu/progressions/

CCSSM was built on progressions that describe how a topic changes across a number of grade levels and are informed both by research on children's cognitive development and by the logical structure of mathematics. The progressions documents are all available online.

Appendix 3
Tasks

This book examines rich tasks that have been used in the classroom to bring to the surface students' understandings and misunderstandings about ratios and proportions. A sampling of these tasks is offered here, in the order in which they appear in the book. At More4U, Appendix 3 includes these tasks and others, formatted for classroom use.

Sands of Time

Jamie solved the following problem for homework:

> Three-fifths of the sand went through a sand timer in 18 minutes. If the rest of the sand goes through at the same rate, how long does it take all the sand to go through the timer?

Jamie got 30 minutes for her answer. Show two different ways that Jamie could have solved the problem.

Pizza Presto: Which Chef Is Faster?

In the Fastest Pizza Maker contest, Keala made three pizzas in five minutes. Casey made four pizzas in nine minutes. If both chefs make pizzas at the same rate as they did in the Fastest Pizza Maker Contest, who do you think will make a pizza faster in a Keala vs. Casey competition?

Measure for Measure, How Long Was That Board?

Jonnine had a board. She cut and used $\frac{2}{5}$ of the board for bracing. She measured the piece used for bracing and found it to be $\frac{3}{4}$ foot long. How long was the original board?

Loads of Luscious Lilikoi

To make Luscious Lilikoi Punch, Austin mixes $\frac{1}{2}$ cup lilikoi passion fruit concentrate with $\frac{2}{3}$ cups water. If he wants to mix concentrate and water in the same ratio to make 28 cups of Luscious Lilikoi Punch, how many cups of lilikoi passion fruit concentrate and how many cups of water will Austin need?

Off the Top of Your Head, Which Is Larger?

As much as possible, use only mental arithmetic to determine which is larger, $\frac{14}{29}$ or $\frac{15}{31}$.

How Many Can You Find?

Find all values of x, $x < 19$, such that $\frac{x}{30}$ can be simplified.

References

Barnett-Clark, Carne, William Fisher, Rick Marks, and Sharon Ross. *Developing Essential Understanding of Rational Numbers for Teaching Mathematics in Grades 3–5.* Essential Understanding Series. Reston, Va.: National Council of Teachers of Mathematics, 2010.

Bryant, Peter, and Terezinha Nunes. *Children's Understanding of Probability: A Literature Review (Full Report).* London, England: Nuffield Foundation, 2012.

Canada, Dan, Mike Gilbert, and Keith Adolphson. "Investigating Mathematical Thinking and Discourse with Ratio Triplets." *Mathematics Teaching in the Middle School* 14 (August 2008): pp. 12–17.

Carter, John A., Gilbert J. Cuevas, Carol Malloy, and Roger Day. *Glencoe Math: Your Common Core Edition.* New York: McGraw-Hill, 2013.

Chval, Kathryn, John Lannin, and Dusty Jones. *Putting Essential Understanding of Fractions into Practice in Grades 3–5.* Putting Essential Understanding into Practice Series. Reston, Va.: National Council of Teachers of Mathematics, 2013.

Cohen, Patricia Cline. "Numeracy in Nineteenth-Century America." In *A History of School Mathematics*, vol. 1, edited by George M. A. Stanic and Jeremy Kilpatrick, pp. 43–76. Reston, Va.: National Council of Teachers of Mathematics, 2003.

Colburn, Warren. *An Arithmetic on the Plan of Pestalozzi, with Some Improvements.* Boston: Cummings, Hilliard, 1821.

——. *Intellectual Arithmetic upon the Inductive Method of Instruction.* Boston: Cummings, Hilliard, 1826.

Cowley, E. B. "Ratio and Proportion in High School Curriculums." *School Science and Mathematics* 107 (December 1937): 1079–88.

Dougherty, Barbara J. "Access to Algebra: A Process Approach." In *The Future of the Teaching and Learning of Algebra*, edited by Helen Chick, Kay Stacey, Jill Vincent, and John Vincent, pp. 207–13. Victoria, Australia: University of Melbourne, 2001.

Grossman, Pamela. *The Making of a Teacher.* New York: Teachers College Press, 1990.

Fishbein, Efraim, Maria Deri, Maria Sainati Nello, and Maria Sciolis Marino. "The Role of Implicit Models in Solving Verbal Problems in Multiplication and Division." *Journal for Research in Mathematics Education* 16 (January 1985): 3–17.

Hackenberg, Amy J., and Erik S. Tillema. "Constructive Resources for Algebraic Reasoning: Middle School Students' Construction of Fraction Composition Schemes." Paper presented at the annual conference of the American Educational Research Association, Montreal, Canada, April 2005.

Hill, Heather C., Brian Rowan, and Deborah Loewenberg Ball. "Effects of Teachers' Mathematical Knowledge for Teaching on Student Achievement." *American Educational Research Journal* 42 (Summer 2005): 371–406.

Kahan, Dan M., Ellen Peters, Erica Cantrell Dawson, and Paul Slovic. "Motivated Numeracy and Enlightened Self-Government." The Cultural Cognition Project at Yale Law School, Working Paper No. 307 (September 2013). http://papers.ssrn.com/sol3/papers.cfm?abstract_id=2319992.

Lamberg, Teruni. *Whole Class Mathematics Discussions: Improving In-Depth Mathematical Thinking and Learning.* Boston: Allyn & Bacon, 2012.

Lamon, Susan J. "Ratio and Proportion: Connecting Content and Children's Thinking." *Journal for Research in Mathematics Education* 24 (January 1993): 41–61.

———. "Ratio and Proportion: Cognitive Foundations in Unitizing and Norming." In *The Development of Multiplicative Reasoning in the Learning of Mathematics*, edited by Guershon Harel and Jere Confrey, pp. 89–120. Albany: State University of New York Press, 1994.

———. *Teaching Fractions and Ratios for Understanding: Essential Content and Instructional Strategies for Teachers.* 1st ed. Mahwah, N.J.: Lawrence Erlbaum, 1999.

———. "Presenting and Representing from Fractions to Rational Numbers." In *The Roles of Representation in School Mathematics,* 2001 Yearbook of the National Council of Teachers of Mathematics (NCTM), edited by Albert A. Cuoco, pp. 146–55. Reston, Va.: NCTM, 2001.

Lesh, Richard, Thomas Post, and Merlyn Behr. "Proportional Reasoning." In *Number Concepts and Operations in the Middle Grades*, edited by James Hiebert and Merlyn Behr, pp. 93–118, Reston, Va.: National Council of Teachers of Mathematics, 1988.

Lo, Jane-Jane, and Tad Watanabe. "Developing Ratio and Proportion Schemes: A Story of a Fifth Grader." *Journal for Research in Mathematics Education* 28 (March 1997): 216–36.

Lobato, Joanne, and Amy B. Ellis. *Developing Essential Understanding of Ratios, Proportions, and Proportional Reasoning for Teaching Mathematics in Grades 6–8.* Essential Understanding Series. Reston, Va.: National Council of Teachers of Mathematics, 2010.

Lobato, Joanne, and Daniel Siebert. "Quantitative Reasoning in a Reconceived View of Transfer." *Journal of Mathematical Behavior* 21, no. 1 (2002): 87–116.

Lobato, Joanne, and Eva Thanheiser. "Developing Understanding of Ratio as Measure as a Foundation for Slope." In *Making Sense of Fractions, Ratios, and Proportions*, 2002 Yearbook of the National Council of Teachers of Mathematics (NCTM), edited by Bonnie Litwiller, pp. 162–75. Reston, Va.: NCTM, 2002.

Magnusson, Shirley, Joseph Krajcik, and Hilda Borko. "Nature, Sources, and Development of Pedagogical Content Knowledge for Science Teaching." In *Examining Pedagogical Content Knowledge*, edited by Julie Gess-Newsome and Norman G. Lederman, pp. 95–132. Dordrecht, The Netherlands: Kluwer Academic, 1999.

Michell, Joel. "The Logic of Measurement: A Realist Overview." *Measurement* 38, no. 4 (2005): 285–94.

National Governors Association Center for Best Practices and Council of Chief State School Officers (NGA Center and CCSSO). *Common Core State Standards for Mathematics. Common Core State Standards (College-and Career-Readiness Standards and K–12 Standards in English Language Arts and Math)*. Washington, D.C.: NGA Center and CCSSO, 2010. http://www.corestandards.org.

Nunes, Terezinha, Despina Desli, and Daniel Bell. "The Development of Children's Understanding of Intensive Quantities." *International Journal of Educational Research* 39 (2003): 651–75.

Olson, Melfried, Fay Zenigami, and Hannah Slovin. "Solving Fraction Worded Problems with the Common Numerator Approach: A Look at Strategies, Reasoning and Models Used by Middle Grades Students." In *6th Annual Hawaii International Conference on Education Conference Proceedings*, pp. 16–29. Honolulu: Hawaii International Conference on Education, 2008.

Olson, Melfried, and Fay Zenigami. "Exploring Equivalent Fractions with the Graphing Calculator." *Mathematics Teaching in the Middle School* 14 (February 2009): 326–29.

Olson, Travis, and Melfried Olson. "The Importance of Context in Presenting Fraction Problems to Help Students Formulate Models and Representations as Solution Strategies." *NCSM Journal of Mathematics Education Leadership* 14 (Fall/Winter 2013): 38–47.

Pike, Nicholas. *A New and Complete System of Arithmetick, Composed for the Use of the Citizens of the United States*. 7th ed. Boston: Thomas & Andrews, 1809.

Popham, W. James. "Defining and Enhancing Formative Assessment." Paper presented at the CCSSO State Collaborative on Assessment and Student Standards FAST meeting, Austin, Tex., October 10–13, 2006.

Shulman, Lee S. "Those Who Understand: Knowledge Growth in Teaching." *Educational Researcher* 15, no. 2 (1986): 4–14.

——. "Knowledge and Teaching." *Harvard Educational Review* 57, no. 1 (1987): 1–22.

Sjostrom, Mary Pat, Melfried Olson, and Travis Olson. "An Examination of the Understanding of Three Groups of Preservice Teachers on Fraction Worded Problems." In *The Proceedings of the 32nd Annual Conference of the North American Chapter of the International Group for the Psychology of Mathematics Education*, edited by Patricia Brosnan, Diana B. Erchick and Lucia Flevares, pp. 1139-47. Columbus: Ohio State University, 2010.

Smith, John P. "The Development of Students' Knowledge of Fractions and Ratios." In *Making Sense of Fractions, Ratios, and Proportions*, 2002 Yearbook of the National Council of Teachers of Mathematics (NCTM), edited by Bonnie Litwiller, pp. 3–18. Reston, Va.: NCTM, 2002.

Smith, Margaret Schwan, and Mary Kay Stein. *5 Practices for Orchestrating Productive Mathematics Discussions*. Reston, Va.: National Council of Teachers of Mathematics, 2011.

Sowder, Judith, Barbara Armstrong, Susan Lamon, Martin Simon, Larry Sowder, and Alba Thompson. "Educating Teachers to Teach Multiplicative Structures in the Middle Grades." *Journal of Mathematics Teacher Education* 1, no. 2 (1998): 127–55.

Steffe, Leslie P. "Children's Multiplying Schemes." In *The Development of Multiplicative Reasoning in the Learning of Mathematics*, edited by Guershon Harel and Jere Confrey, pp. 3–39. Albany: State University of New York Press, 1994.

Thompson, Patrick W. "The Development of the Concept of Speed and Its Relationship to Concepts of Rate." In *The Development of Multiplicative Reasoning in the Learning of Mathematics*, edited by Guershon Harel and Jere Confrey, pp. 179–234. Albany: State University of New York Press, 1994.

Thompson, Patrick W., and Luis A. Saldanha. "Fractions and Multiplicative Reasoning." In *A Research Companion to "Principles and Standards for School Mathematics,"* edited by Jeremy Kilpatrick, W. Gary Martin, and Deborah Schifter, pp. 95–113. Reston, Va.: National Council of Teachers of Mathematics, 2003.

Wiliam, Dylan. "Keeping Learning on Track: Classroom Assessment and the Regulation of Learning." In *Second Handbook of Research on Mathematics Teaching and Learning*, edited by Frank K. Lester, Jr., pp. 1053–1098. Charlotte, N.C.: Information Age; Reston, Va.: National Council of Teachers of Mathematics, 2007.

Wu, Hung-Hsi. "What's Sophisticated about Elementary Mathematics? Plenty—That's Why Elementary Schools Need Math Teachers." *American Educator* (Fall 2009): 4–14.

———. "Pre-Algebra [Draft]." http://math.berkeley.edu/%7Ewu/Pre-Algebra.pdf, November 2011.

Yinger, Robert J. "The Conversation of Teaching: Patterns of Explanation in Mathematics Lessons." Paper presented at the meeting of the International Study Association on Teacher Thinking, Nottingham, England, May 1998.